EVOLUTION OF LIFE

EVOLUTION OF LIFE

CECIE STARR / RALPH TAGGART

Christine A. Evers

Lisa Starr

BIOLOGY
THE UNITY AND DIVERSITY OF LIFE
ELEVENTH EDITION

THOMSON

BROOKS/COLE

Australia • Brazil • Canada • Mexico • Singapore
Spain • United Kingdom • United States

PUBLISHER Jack C. Carey

VICE-PRESIDENT, EDITOR-IN-CHIEF Michelle Julet

SENIOR DEVELOPMENT EDITOR Peggy Williams

ASSOCIATE DEVELOPMENT EDITOR Suzannah Alexander

EDITORIAL ASSISTANT Chris Ziemba, Kristina Razmara

TECHNOLOGY PROJECT MANAGER Keli Amann

MARKETING MANAGER Ann Caven

MARKETING ASSISTANT Brian Smith

MARKETING COMMUNICATIONS MANAGER Nathaniel
  Bergson-Michelson

PROJECT MANAGER, EDITORIAL PRODUCTION Andy Marinkovich

CREATIVE DIRECTOR Rob Hugel

PRINT BUYER Karen Hunt

PERMISSIONS EDITOR Joohee Lee, Sarah Harkrader

PRODUCTION SERVICE Grace Davidson & Associates

TEXT AND COVER DESIGN Gary Head

PHOTO RESEARCHER Myrna Engler

COPY EDITOR Kathleen Deselle

ILLUSTRATORS Gary Head, ScEYEnce Studios, and Lisa Starr

COMPOSITOR Lachina Publishing Services

TEXT AND COVER PRINTER R.R. Donnelley/Willard

COVER IMAGE Ko'olau Mountains on the windward side of
Oahu, part of the Hawaiian Archipelago that is a natural
laboratory for the study of evolution.
Photographer Russ Lowgren

Printed in the United States of America
1 2 3 4 5 6 7 09 08 07 06 05

Library of Congress Control Number: 2005931273

ISBN 0-495-12579-2

For more information about our products, contact us at:
Thomson Learning Academic Resource Center
1-800-423-0563

For permission to use material from this text or product, submit
a request online at http://www.thomsonrights.com.
Any additional questions about permissions can be submitted
by e-mail to thomsonrights@thomson.com.

BOOKS IN THE BROOKS/COLE BIOLOGY SERIES

Biology: The Unity and Diversity of Life, Eleventh, Starr/Taggart
Engage Online for Biology: The Unity and Diversity of Life
Biology: Concepts and Applications, Sixth, Starr
Basic Concepts in Biology, Sixth, Starr
Biology Today and Tomorrow, Second, Starr
Biology, Seventh, Solomon/Berg/Martin
Human Biology, Sixth, Starr/McMillan
Biology: A Human Emphasis, Sixth, Starr
Human Physiology, Fifth, Sherwood
Fundamentals of Physiology, Second, Sherwood
Human Physiology, Fourth, Rhoades/Pflanzer

Laboratory Manual for Biology, Fourth, Perry/Morton/Perry
Laboratory Manual for Human Biology, Morton/Perry/Perry
Photo Atlas for Biology, Perry/Morton
Photo Atlas for Anatomy and Physiology, Morton/Perry
Photo Atlas for Botany, Perry/Morton
Virtual Biology Laboratory, Beneski/Waber
Introduction to Cell and Molecular Biology, Wolfe
Molecular and Cellular Biology, Wolfe
Biotechnology: An Introduction, Second, Barnum

Introduction to Microbiology, Third, Ingraham/Ingraham
Microbiology: An Introduction, Batzing
Genetics: The Continuity of Life, Fairbanks/Anderson
Human Heredity, Seventh, Cummings
Current Perspectives in Genetics, Second, Cummings
Gene Discovery Lab, Benfey

Animal Physiology, Sherwood, Kleindorf, Yarcey
Invertebrate Zoology, Seventh, Ruppert/Fox/Barnes
Mammalogy, Fourth, Vaughan/Ryan/Czaplewski
Biology of Fishes, Second, Bond
Vertebrate Dissection, Ninth, Homberger/Walker

Plant Biology, Second, Rost/Barbour/Stocking/Murphy
Plant Physiology, Fourth, Salisbury/Ross
Introductory Botany, Berg

General Ecology, Second, Krohne
Essentials of Ecology, Third, Miller
Terrestrial Ecosystems, Second, Aber/Melillo
Living in the Environment, Fourteenth, Miller
Environmental Science, Tenth, Miller
Sustaining the Earth, Seventh, Miller
Case Studies in Environmental Science, Second, Underwood
Environmental Ethics, Third, Des Jardins
Watersheds 3—Ten Cases in Environmental Ethics, Third,
Newton/Dillingham

Problem-Based Learning Activities for General Biology, Allen/Duch
The Pocket Guide to Critical Thinking, Second, Epstein

WebTutor™ © 2006 Thomson Learning, Inc. All Rights Reserved. Thomson
Learning WebTutor™ is a trademark of Thomson Learning, Inc.

*Due to contractual restrictions, Thomson can offer vMentor only to
institutions of higher education (including post-secondary, proprietary
schools) within the United States. We are unable to offer it outside the
US or to any other US domestic customers.*

Thomson Higher Education
10 Davis Drive
Belmont, CA 94002-3098
USA

Asia (including India)
Thomson Learning
5 Shenton Way
#01-01 UIC Building
Singapore 068808

Australia/New Zealand
Thomson Learning Australia
102 Dodds Street
Southbank, Victoria 3006
Australia

Canada
Thomson Nelson
1120 Birchmount Road
Toronto, Ontario M1K 5G4

UK/Europe/Middle East/Africa
Thomson Learning
High Holborn House
50/51 Bedford Row
London WC1R 4LR
United Kingdom

CONTENTS IN BRIEF

Highlighted chapters are not included in Evolution of Life

DETAILED CONTENTS

PREFACE

Biology's overriding paradigm is one of interpreting life's spectacular diversity as having evolved from simple molecular beginnings. For decades, researchers have been chipping away at the structural secrets of biological molecules. They are showing how different kinds are put together, how they function, and what happens when they mutate. They are casting light on how life originated, what happened over the past 3.8 billion years, and what the future may hold for humans and other organisms, individually and collectively.

This is profound stuff. *Yet introductory textbooks—including previous editions of this one—have not given students enough information to interpret for themselves the remarkable connection between molecular change, evolution, and their own lives.* We continue to rebuild this book in ways that clarify the connection. When students "get" the big picture of life, they become confident enough to think critically about the past, present, and future on their own. This is one of biology's greatest gifts.

MAKE THE DIRECT CONNECTION

New to this edition are connections that have direct and often controversial impact on our lives. What students learn today may help them make decisions tomorrow, in the voting booth as well as in their personal lives.

How Would You Vote?   Each chapter starts with an essay on a current issue that relates to its content (see list at right). Essays are expanded in custom *videoclips* and exercises on the student website and in the Power-Lecture, a one-stop PowerPoint tool for instructors.

We ask students in each chapter, *How would you vote on research or an application related to this issue?* Then the exercise invites them to read a balanced selection of articles, pro and con, before voting. Students throughout the country are already voting *online*, and are accessing campuswide, statewide, and nationwide tallies. This interactive approach to issues reinforces the premise that individual actions can make a difference.

HELP STUDENTS TRACK CONNECTIONS

Linking Concepts   Also new to this edition are links to concepts within and between chapters. Students have an easier time making connections when links guide them. Each chapter's opening spread has a section-by-section list of *key concepts*, each with a simple title. We repeat the titles at the top of appropriate text pages as reminders of the conceptual organization. A brief list of *links to earlier concepts* helps students assess concepts they should already know before they begin. For instance, before reading about neural function, they may wish to scan an earlier chapter's section on active transport. We repeat the linking icons in page margins to help students refer back to relevant sections.

Concept Spreads   Judging from extensive feedback, *concept spreads* are immensely popular. Students report that they can easily study and master parts of a chapter

IMPACTS/ISSUES ESSAYS

when they have a small window of time rather than wading into an overwhelming mass of information. We place all text, art, and evidence in support of each concept in one section that consists of two facing pages, at most. Each section starts at a numbered tab and ends with a blue-boxed *on-page summary*, which students can use to check whether they understand the main points before turning to the next concept spread.

Concept spreads offer teachers flexibility in assigning topics to fit course requirements. Those who spend less time on, say, photosynthesis might bypass details on the properties of light and ATP formation. All spreads are part of the chapter topic, but some offer added depth.

Read-Me-First Art   *Read-me-first diagrams* are visual learning devices. Students can walk through them step-by-step as a preview of the text. The same art occurs in narrated, animated form online at BiologyNow and on the PowerLecture DVD. Repeated exposure to material in a variety of modalities accommodates diverse learning styles and reinforces knowledge of concepts.

Critical Thinking   Like all textbooks at this level, we walk students through examples of problem solving and experiments throughout the book. We integrate many simple experiments in the text proper. *Focus on Science* essays highlight the detailed ones. The entries *Experiment, examples*, and *Test, examples* in the main index list our selections.

Each chapter ends with a number of *Critical Thinking* questions, some with art, and some more challenging than others. They invite students to think outside the memorization box. We include an *annotated scientific paper* as an example of how scientists work and then report their findings (Appendix IX).

We selectively use some chapter introductions, even a few whole chapters, to emphasize scientific methods. However, introductory students are not scientists, so we do not shoehorn *all* material into an experimental context. We cannot expect them to learn the language and processes of science by intuition alone; we help them build these skills in a paced way.

Connections essays, flagged by vertical red bands on page edges, invite students to step back and connect the dots between chapters and units. An *Epilogue* invites them to make the most sweeping connection of all.

WRITE CLEARLY AND SELECTIVELY

Students often complain that their textbooks are dry and boring. We write to engage them without glossing over the science, even through such formidable topics as the laws of thermodynamics (Section 6.1).

To keep book length manageable, we were selective about which topics to include so that we could allocate enough space to explain them clearly. For instance, most nonmajors simply do not want to memorize each catalytic step of crassulacean acid metabolism. They do want to learn about the basis of sex, and many women would like to know what will go on inside their body if they become pregnant. Sufficient information will help them make informed decisions on many biology-related

issues, such as STDs, fertility drugs, prenatal diagnoses, gene therapies, and stem cell research. Over the years, students have written to tell us that they did read such material closely even when it was not assigned.

Our choices for which topics to condense, expand, or delete reflect more than three decades of feedback from teachers of millions of students around the world.

OFFER STUDENTS EASY-TO-USE MEDIA TOOLS

With this edition, logging in to online assets is simpler than before. Students register their free **1pass access code** at http://1pass.thomson.com and then log in to access all resources outlined below. An access code is packaged in each new copy of the book and also is available through e-commerce.

BiologyNow™   For each text chapter, *BiologyNow* provides *animated tutorials* as well as guidance to other resources that can help students master the content. Responses to *diagnostic pretest questions* yield a student-driven, *personalized learning plan*. Answer a question incorrectly, and the plan lists text sections, figures, and chapter videos. The plan also gives links to relevant animation.

The *How Do I Prepare* section has tutorials in math, chemistry, graphing, and basic study skills. *vMentor* is an online service with a live tutor who interacts with students through voice communication, whiteboard, and text messaging.

The *post-test* can be used as self-assessment tool or submitted to an instructor. BiologyNow is built in iLrn, Thomson Learning's course management system, so that answers and results can be fed directly to an electronic gradebook. BiologyNow also can be integrated with WebCT and Blackboard, so students may log in through these systems as well.

InfoTrac College Edition™   1pass also grants access to *InfoTrac College Edition*, an online database with more than 4,000 periodicals and close to 18 million articles. The articles are in full-text form and can be located easily and quickly with a *key word* search.

In BiologyNow, the *How Would You Vote* exercise for each chapter references specific InfoTrac articles and websites. As students vote on each issue, the site provides a running tally by campus, state, and the country. Instructors can assign the exercises from iLrn, or students can access them through the free website at:

http://biology.brookscole.com/starr11

This website also features more InfoTrac articles and web links, as well as *interactive flashcards* that define all of the book's boldface terms with pronunciation guides.

Also, an online *Issues and Resources Integrator* correlates chapter sections with applications, videos, InfoTrac articles, and websites. This guide is updated each semester.

Audiobook   New college editions of the textbook can also include free access to the *Audiobook*. Students can either listen to narration online or download mp3 files for use on a portable mp3 player.

Previous adopters of *Biology: The Unity and Diversity of Life* who continue to use the book may wish to scan this list of refinements and updates in chapter content. Again, all chapters start with new essays on current issues that are more likely to engage students.

A New Tree of Life Thanks to the wealth of information from biochemistry and molecular biology, the tree of life is coming into sharp focus. Working closely with Walter Judd at the University of Florida, we became confident enough to create a simple but powerful graphic of the evolutionary connections, including the main archaean and protist groups. We present the tree in Section 19.7 as a prelude to the diversity unit. We repeat parts of it in the diversity chapters, as regional road maps. The first chapter does have a simplified classification system in which six kingdoms are subsumed into three domains. It is all that students will require until they reach the later units on evolutionary principles and biodiversity.

We use informal names in the tree and in much of the text. Our advisors and focus-group participants agree that it is more important for introductory students to think about evolutionary connections than ongoing refinements in taxonomy. For reference purposes, Appendix I still provides taxonomic names for major groups.

Energy and Life's Organization Throughout the book, we strengthened the concept of energy flow as the basis of life's levels of organization. Students generally do not find bioenergetics an endearing topic, but we have two new, nonthreatening sections that might engage them through the use of real-life examples (6.1, 6.2). This conceptual overview may help them make sense of metabolism. The photosynthesis chapter (7) has reorganized, streamlined text, new art (including an update on chloroplast structure), and good bioenergetic and evolutionary threads.

Evolution We continue to apply evolutionary theory all through the book. The early, simple overview in Section 1.4 is enough to get students thinking about the concept without bogging them down in details. As examples, we invite them to think about why sexual reproduction evolved (the new Chapter 10 introduction), a possible evolutionary connection between mitosis and meiosis (10.6), and past and potential future threats of methane hydrate deposits on the seafloor. We continue to expand on the connections between mutation, evolution, and development, or "evo-devo" (15.3, 17.8, 17.9, 18.1, 25.1, 25.2, 25.8, 26.4, 26.14, 32.6, 34.1, 39.1, and 43.5 are some examples). The macroevolution chapter (18) has a new essay and text section on the adaptive radiation and current extinctions of the Hawaiian honeycreepers.

Cell Communication We strengthened the art and text presentations of signaling pathways. Sections 22.12 and 25.1 can get students thinking about the evolutionary origins of cell communication. Section 28.5 introduces a generalized graphic of signal reception, transduction, and response that we repeat throughout the animal and plant anatomy units. Section 28.5 also has a specific example: apoptosis in the development of the human hand. We build on this introduction in the sections on neural function (34.1–34.5), sensory function (35.1), hormone action in animals (36.2) and in plants (32.3), immune function (39.4, 39.6, 39.7), urine formation (42.4), and embryonic development (43.2, 43.5).

Unit I Chapter 1: reorganized, simplified, with a new section on experimental tests. Chapter 2: more accessible introduction to electrons and energy levels. Chapter 3: revised sections on levels of protein structure. Chapter 4: new art and text on cell structures. Chapter 5: sections on diffusion, transport mechanisms, and osmosis revised for clarity. Chapter 6: new art, more accessible writing. Chapter 7: extensively revised, updated. Chapter 8: new art for the Krebs cycle and the electron transfer system.

Unit II Chapter 9: new introduction and new essay on cancer. Chapter 10: stronger evolutionary framework for explaining meiosis. Chapter 11: section on impact of crossing over on gene linkage moved to this chapter; new genetics problems on rose and watermelon hybrid experiments. Chapter 12: reorganization of sections on autosomal and X-linked inheritance. Chapter 13: new, updated essay on how cloning methods manipulate information in DNA. Chapter 14: subtle rewrites for clarity, updated alternative mRNA processing. Chapter 15: reorganized, starts with overview of control points; introduces *ABC* flowering model; new essay on effect of mutant *Drosophila* master genes on development. Chapter 16: major reorganization and rewrite, new sections on genomics, organ factories, and other research that will have enormous impact on our lives.

Unit III Chapter 17: now starts with an essay on mass extinctions as one measure of geologic time; introduces evidence for evolution (fossils, radiometric dating, biogeography and plate tectonics, and comparative morphology, embryology, and biochemistry); has new examples of evolutionary constraints on developmental patterns for plants and animals. Chapter 18: some new examples, as in introductory essay; revised section on genetic drift; ends with essay on adaptation (adaptation to what?). Chapter 19: revised text and art, some new examples for reproductive isolating mechanisms and speciation models; updated sections on taxonomy and systematics; new tree of life.

Unit IV Chapter 20: now a less detailed introduction and timeline for the diversity chapters; updated sections on origin of life and of organelles. Chapter 21: essay on West Nile virus; rewrite and updates on bacterial and archaean lineages; updated essay on infectious disease, including SARS and prion diseases. Chapter 22: basically a new chapter on "protists" with clarified phylogenies; has new essay on amoebozoans and the origin of cell communication and multicellularity. Chapter 23: starts and ends with essays on deforestation; new overview of plant origins and evolution; charophytes now included in plant clade; more on flowering plants and pollinators to set the stage for Section 32.3. Chapter 24: new essay on lichens; new material on chytrids, microsporidians, *Neurospora* life cycle, and endophytic fungi.

Chapter 25: major new, updated sections on animal origins and characteristics, including the basis of large, internally complex bodies; reordering of placozoans, flatworms, roundworms, other major groups; essay on cephalopod evolution. Chapter 26: *Archaeopteryx* essay; revised sections on chordate family tree and vertebrate evolutionary trends; aquatic origin of tetrapods with new art; focus essay on vanishing amphibians; sections on dinosaurs now integrated in this chapter; revised text and art on rise of amniotes; new text and art on adaptive radiation of mammals; updated text and art on primate and hominid evolutionary trends.

Chapter 27: now includes a section on environmental impact of human population growth; Rachel Carson essay; conservation biology and sustaining biodiversity, with Monteverde Cloud Forest as an example; ends on a more optimistic note with examples of what might be accomplished by thinking outside the box.

Introduction to Units V and VI New to this edition. Chapter 28: presents common challenges facing plants and animals at increasingly complex levels structural organization; introduces concept of homeostasis and requirements for gas exchange, internal transport, and cell communication, as listed earlier; animal examples of homeostasis are temperature regulation and oxygen deficiency at high altitudes; plant examples are systemic acquired resistance and leaf folding in response to shifts in environmental conditions.

Unit V Chapter 29: reorganized, rewritten introduction to plant tissues with new graphics and micrographs. Chapter 31: incorporates new essay on coevolution of flowering plants and pollinators, including text and photograph of the hawkmoth and the Christmas Star; additional diagrams, improved classification of fruits. Chapter 32: heavily revised, updated; auxin signaling pathway used as new example of cell communication; ends the unit with a new connections essay on global protein deficiency and quinoa research (earlier edition's essay on pesticides is now in Chapter 47 as part of an expanded essay on biomagnification).

Unit VI Chapter 33: new essay on stem cell research; new micrographs; tissue descriptions rewritten, a bit more on muscle tissue; chapter builds on Section 25.1 introduction to compartmentalization (division of labor) as emergent property of multicellularity; new essay on vertebrate skin. Chapter 34: all sections simplified and reorganized; opens with Ecstasy essay; new overview of invertebrate and central/peripheral vertebrate nervous systems; rewritten sections on neurotransmitters and neuromodulators; new section on neuroglia; updated essay on psychoactive drugs. Chapter 35: new essay on oceanic noise pollution. Chapter 36: new essay and text

on hormone disruptors, amphibians. and human sperm counts; major text reorganization (gland by gland); new essay on diabetes and impact of stress on health.

Chapter 37: new section on origin of invertebrate and vertebrate skeletons. Chapter 38: new text, art on cardiac conduction system. Chapter 39: fully revised chapter; more accessible, yet reflects current research and up-to-date concepts, including evolutionary and functional interrelationship between adaptive and innate immunity; easier-to-follow art; update on HIV/AIDS. Chapter 40: rewrite on countercurrent flow in fish gills; new text on controls over respiration. Chapter 41: opens with essay on hormones and appetite; updated human nutritional requirements. Chapter 42: major reorganization and new step-by-step graphics and text on urine formation; expanded emphasis on kidney disorders.

Chapter 43: frog life cycle now integrated with the overview of stages of development; sections on cleavage, pattern formation, and constraints on development have been clarified; new Critical Thinking question on RNA interference for interested students (page 769); updates on aging hypotheses. Chapter 44: text tightened with some reorganization (sections on fertility control and STDs now located after human reproduction and before development); updates on STDs, pregnancy guidelines, fetal alcohol syndrome; oxytocin secretion during labor a new example of positive feedback; essay on bioethics of interventions in fertility now at end of chapter.

Unit VII Chapter 45: unit now opens with essay on reindeer overshooting carrying capacity on St. Matthews Island; revised sections on limiting factors; updates on human population growth; demographic transition model now qualified. Chapter 46: rewrites and new art for mutualism, competitive interactions, predation models; new examples of parasitism; revised community structure models; updates on *Codium* aquarium strain, kudzu, rabbits in Australia; some additions to section on biogeographic patterns; Critical Thinking question on Wallace's line (page 841). Note that Chapter 50 sections of the previous edition are now rolled into Chapters 47 and 48. Chapter 47: global warming and bayous; revised essays on pesticides and biomagnification, energy flow, watershed experiments; global water crisis integrated in this chapter; carbon cycle updates; global warming updates; revised nitrogen cycle. Chapter 48: text on solar energy and wind energy now integrated in section on global air circulation patterns; essay on air pollution (ozone thinning, smog, acid rain, particulates); better historical look at biogeographic realms; new section on differences in sunlight, soils, and moisture, with desert soils and moisture differences as case study; new essay on desertification, including a model for feedback loops between desertification, climate change, and losses in biodiversity; expansion of freshwater provinces with a section on water pollution; new section on north central Pacific gyre as garbage dump. Chapter 49: update on genetics of courtship behavior in fruit flies and pair bonds in voles; new sections on primate social behavior and on biological aspects of human behavior.

ACKNOWLEDGMENTS

Once again I am thinking about all of the advisors and reviewers who helped shape the content of this book in significant ways. Bits and pieces of earlier critiques still flit through my mind, and they are making me realize that educators are symbionts of the highest order. Why else would they give so much of themselves, year after year, to the cooperative enterprise called education? The demands are great and the money is not. Somehow they all have become hardwired to contribute to the greater good. Each year when I interact with them, some new manifestation of their commitment manages to take my breath away; which is why I am always uncomfortable with calling this book "the Starr book." It is our book.

Dan Fairbanks, John Jackson, Walt Judd, and Bill Wischusen have been fearless about approaching this established book with a sledgehammer. They are aware that research directions change, students change, and textbooks must change with them. They dig out flaws in my thinking with needles. I admire them immensely.

Biology: The Unity and Diversity of Life wins awards for subsuming design and art in the service of pedagogy instead of slapping them on as afterthoughts. All it takes is abiding passion for biology and a compulsive urge to assess trends in education, keep up with research, write, create art, and finely lay out workable concept spreads, all at the same time. For some time now, I have been mentoring Lisa Starr and Christine Evers in the odd science of being multitasking authors. Lisa is educated in biochemistry, biotechnology, and computer graphics; Chris in animal behavior, evolutionary biology, and media technologies. With their combined talents and dedication, they have become indispensable partners.

Starting with Susan Badger, Michael Johnson, and Sean Wakely, Thomson Learning proved once again with this edition that it is one of the world's foremost publishers. Michelle Julet, what would I do without your laser focus, arm twisting, intelligence, and wit? Peggy Williams is Empress of the Team. So many of the improvements in this book are direct outcomes of her extraordinary talent for organizing and interpreting the results of class tests, workshops, and focus groups. Andy Marinkovich and dear Grace Davidson continue to take my world-class compulsivity in stride and still get the book out on time. Amazing. Gary Head, aka The Rock, suffers through being my designer, year in, year out. Thank goodness for Myrna Engler's devotion to the team and Suzannah Alexander's quiet competence. Ann Caven, Kathie Head, Laura Argento, Keli Amann, Marlene Veach, Karen Hunt, Chris Ziemba, Kristina Razmara, the professionals at Lachina—the list goes on, but no listing conveys how this team interacts to create something extraordinary.

Jack Carey, how long ago was it that you signed me to do eighteen books to keep me from writing for anybody else in this lifetime? I never did get around to the other fourteen, but probably you forgive me. Through our long partnership, we helped move textbook publishing in new directions, to the benefit of students around the world. Never would have done it without you, partner.

CECIE STARR, *November 2005*

MAJOR ADVISORS AND REVIEWERS

DANIEL FAIRBANKS *Brigham Young University*
JOHN D. JACKSON *North Hennepin Community College*
WALTER JUDD *University of Florida*
E. WILLIAM WISCHUSEN *Louisiana State University*

JOHN ALCOCK *Arizona State University*
CHARLOTTE BORGESON *University of Nevada*
DEBORAH C. CLARK *Middle Tennessee University*
MELANIE DEVORE *Georgia College and State University*
TOM GARRISON *Orange Coast College*
DAVID GOODIN *The Scripps Research Institute*
CHRISTOPHER GREGG *Louisiana State University*
PAUL E. HERTZ *Barnard College*
TIMOTHY JOHNSTON *Murray State University*
EUGENE KOZLOFF *University of Washington*
ELIZABETH LANDECKER-MOORE *Rowan University*
KAREN MESSLEY *Rock Valley College*
THOMAS L. ROST *University of California, Davis*
LAURALEE SHERWOOD *West Virginia University*
STEPHEN L. WOLFE *University of California, Davis*

CONTRIBUTORS: 2004–2005 WORKSHOPS AND REVIEWS

ANDERSON, DIANNE *San Diego City College*
ASMUS, STEVE *Centre College*
BLAUSTEIN, ANDREW R. *Oregon State University*
BROSSAY, LAURENT *Brown University*
CAPORALE, DIANE *University of Wisconsin-Stevens Point*
CAWTHORN, J. MICHELLE *Georgia Southern University*
COLLIER, ALEXANDER *Armstrong Atlantic State University*
CONWAY, ARTHUR F. *Randolph-Macon College*
DAVIS, GEORGE T. *Bloomsburg University*
DEGRAUW, EDWARD *Portland Community College*
DELANEY, CYNTHIA LEIGH *University of South Alabama*
DESAIX, JEAN *University of North Carolina*
DIERINGER, GREG *Northwest Missouri State University*
D'ORGEIX, CHRISTIAN *Virginia State University*
D'SILVA, JOSEPH G. *Norfolk State University*
FONG, APRIL *Portland Community College*
GARCIA, RIC A. *Clemson University*
HAIGH, GALE *McNeese State University*
HEARRON, MARY *Northeast Texas Community College*
HENDERSON, WILEY J. *Alabama A&M University*
HINTON, JULIANA *McNeese State University*
HUANG, SARA *Los Angeles Valley College*
HUFFMAN, DONNA *Calhoun Community College*
JONES, KEN *Dyersburg State Community College*
JULIAN, GLENNIS *University of Texas at Austin*
KROLL, WILLIAM *Loyola University, Chicago*
KURDZIEL, JOSEPHINE *University of Michigan*
LAMMERT, JOHN M. *Gustavus Adolphus College*
LAZOTTE, PAULINE *Valencia Community College*
MADTES, JR., PAUL *Mount Vernon Nazarene University*
MAJDI, BERNARD *Waycross College*
MATA, JUAN LUIS *University of Tennessee at Martin*
MCKEAN, HEATHER R. *Eastern Washington University*
METZ, TIMOTHY *Campbell University*
MOSS, ANTHONY *Auburn University*
NEUFELD, HOWARD S. *Appalachian State University*
NOLD, STEPHEN *University of Wisconsin*
ORR, CLIFTON *University of Arkansas at Pine Bluff*
PETERS, JOHN *College of Charleston*
PLUNKETT, JENNIE *San Jacinto College*
POMARICO, STEVEN M. *Louisiana State University*
RAINES, KIRSTEN *San Jacinto College*
REED, ROBERT N. *Southern Utah University*
RIBEIRO, WENDA *Thomas Nelson Community College*
RICHEY, MARGARET G. *Centre College*
ROBERTS, LAUREL *University of Pittsburgh*

ROMANO, FRANK A., III, *Jacksonville State University*
RUPPERT, ETTA *Clemson University*
SHOFNER, MARCIA *University of Maryland, College Park*
SIEVERT, GREG *Emporia State University*
SIMS, THOMAS L. *Northern Illinois University*
SONGER, STEPHANIE R. *Concord University*
SPRENKLE, AMY B. *Salem State College*
ST. CLAIR, LARRY *Brigham Young University*
TEMPLET, ALICE *Nicholls State University*
TURELL, MARSHA *Houston Community College*
WALSH, PAT *University of Delaware*
WILKINS, HEATHER DAWN *University of Tennessee at Martin*
WINDELSPECHT, MICHAEL *Appalachian State University*
WYGODA, MARK *McNeese State University*
ZAHN, MARTIN D. *Thomas Nelson Community College*
ZANIN, KATHY *The Citadel*

CONTRIBUTORS: INFLUENTIAL CLASS TESTS AND REVIEWS

ADAMS, DARYL *Minnesota State University, Mankato*
ANDERSON, DENNIS *Oklahoma City Community College*
BENDER, KRISTEN *California State University, Long Beach*
BOGGS, LISA *Southwestern Oklahoma State University*
BORGESON, CHARLOTTE *University of Nevada*
BOWER, SUSAN *Pasadena City College*
BOYD, KIMBERLY *Cabrini College*
BRICKMAN, PEGGY *University of Georgia*
BROWN, EVERT *Casper College*
BRYAN, DAVID W. *Cincinnati State College*
BURNETT, STEPHEN *Clayton College*
BUSS, WARREN *University of Northern Colorado*
CARTWRIGHT, PAULYN *University of Kansas*
CASE, TED *University of California, San Diego*
COLAVITO, MARY *Santa Monica College*
COOK, JERRY L. *Sam Houston State University*
DAVIS, JERRY *University of Wisconsin, LaCrosse*
DENGLER, NANCY *University of California, Davis*
DeSAIX, JEAN *University of North Carolina*
DiBARTOLOMEIS, SUSAN *Millersville University of Pennsylvania*
DIEHL, FRED *University of Virginia*
DONALD-WHITNEY, CATHY *Collin County Community College*
DUWEL, PHILIP *University of South Carolina, Columbia*
EAKIN, DAVID *Eastern Kentucky University*
EBBS, STEPHEN *Southern Illinois University*
EDLIN, GORDON *University of Hawaii, Manoa*
ENDLER, JOHN *University of California, Santa Barbara*
ERWIN, CINDY *City College of San Francisco*
FOREMAN, KATHERINE *Moraine Valley Community College*
FOX, P. MICHAEL *SUNY College at Brockport*
GIBLIN, TARA *Stephens College*
GILLS, RICK *University of Wisconsin, La Crosse*
GREENE, CURTIS *Wayne State University*
GREGG, KATHERINE *West Virginia Wesleyan College*
HARLEY, JOHN *Eastern Kentucky University*
HARRIS, JAMES *Utah Valley Community College*
HELGESON, JEAN *Collin County Community College*
HESS, WILFORD M. *Brigham Young University*
HOUTMAN, ANNE *Cal State, Fullerton*
HUFFMAN, DAVID *Southwestern Texas University*
HUFFMAN, DONNA *Calhoun Community College*
INEICHER, GEORGIA *Hinds Community College*
JOHNSTON, TAYLOR *Michigan State University*
JUILLERAT, FLORENCE *Indiana University, Purdue University*
KENDRICK, BRYCE *University of Waterloo*
KETELES, KRISTEN *University of Central Arkansas*
KIRKPATRICK, LEE A. *Glendale Community College*
KREBS, CHARLES *University of British Columbia*
LANZA, JANET *University of Arkansas, Little Rock*
LEICHT, BRENDA *University of Iowa*
LOHMEIER, LYNNE *Mississippi Gulf Coast Community College*
LORING, DAVID *Johnson County Community College*
MACKLIN, MONICA *Northeastern State University*

MANN, ALAN *University of Pennsylvania*
MARTIN, KATHY *Central Connecticut State University*
MARTIN, TERRY *Kishwaukee College*
MASON, ROY B. *Mount San Jacinto College*
MATTHEWS, ROBERT *University of Georgia*
MAXWELL, JOYCE *California State University, Northridge*
McCLURE, JERRY *Miami University*
McNABB, ANN *Virginia Polytechnic Institute and State University*
MEIERS, SUSAN *Western Illinois University*
MEYER, DWIGHT H. *Queensborough Community College*
MICKLE, JAMES *North Carolina State University*
MILLER, G. TYLER *Wilmington, North Carolina*
MINOR, CHRISTINE V. *Clemson University*
MITCHELL, DENNIS M. *Troy University*
MONCAYO, ABELARDO C. *Ohio Northern University*
MOORE, IGNACIO *Virginia Tech*
MORRISON-SHETTLER, ALLISON *Georgia State University*
MORTON, DAVID *Frostburg State University*
NELSON, RILEY *Brigham Young University*
NICKLES, JON R. *University of Alaska, Anchorage*
NOLD, STEPHEN *University of Wisconsin- Stout*
PADGETT, DONALD *Bridgewater State College*
PENCOE, NANCY *State University of West Georgia*
PERRY, JAMES *University of Wisconsin, Center Fox Valley*
PITOCCHELLI, DR. JAY *Saint Anselm College*
PLETT, HAROLD *Fullerton College*
POLCYN, DAVID M. *California State University, San Bernardino*
PURCELL, JERRY *San Antonio College*
REID, BRUCE *Kean College of New Jersey*
RENFROE, MICHAEL *James Madison University*
REZNICK, DAVID *California State University, Fullerton*
RICKETT, JOHN *University of Arkansas, Little Rock*
ROHN, TROY *Boise State University*
ROIG, MATTIE *Broward Community College*
ROSE, GRIEG *West Valley College*
SANDIFORD, SHAMILI A. *College of Du Page*
SCHREIBER, FRED *California State University, Fresno*
SELLERS, LARRY *Louisiana Tech University*
SHAPIRO, HARRIET *San Diego State University*
SHONTZ, NANCY *Grand Valley State University*
SHOPPER, MARILYN *Johnson County Community College*
SIEMENS, DAVID *Black Hills State University*
SMITH, BRIAN *Black Hills State University*
SMITH, JERRY *St. Petersburg Junior College, Clearwater Campus*
STEINERT, KATHLEEN *Bellevue Community College*
SUMMERS, GERALD *University of Missouri*
SUNDBERG, MARSHALL D. *Emporia State University*
SVENSSON, PETER *West Valley College*
SWANSON, ROBERT *North Hennepin Community College*
SWEET, SAMUEL *University of California, Santa Barbara*
SZYMCZAK, LARRY J. *Chicago State University*
TAYLOR, JANE *Northern Virginia Community College*
TERHUNE, JERRY *Jefferson Community College, University of Kentucky*
TIZARD, IAN *Texas A&M University*
TRAYLER, BILL *California State University at Fresno*
TROUT, RICHARD E. *Oklahoma City Community College*
TURELL, MARSHA *Houston Community College*
TYSER, ROBIN *University of Wisconsin, LaCrosse*
VAJRAVELU, RANI *University of Central Florida*
VANDERGAST, AMY *San Diego State University*
VERHEY, STEVEN *Central Washington University*
VICKERS, TANYA *University of Utah*
VOGEL, THOMAS *Western Illinois University*
WARNER, MARGARET *Purdue University*
WEBB, JACQUELINE F. *Villanova University*
WELCH, NICOLE TURRILL *Middle Tennessee State University*
WELKIE, GEORGE W. *Utah State University*
WENDEROTH, MARY PAT *University of Washington*
WINICUR, SANDRA *Indiana University, South Bend*
WOLFE, LORNE *Georgia Southern University*
YONENAKA, SHANNA *San Francisco State University*
ZAYAITZ, ANNE *Kutztown University of Pennsylvania*

Introduction

Current configurations of the Earth's oceans and land masses—the geologic stage upon which life's drama continues to unfold. This composite satellite image reveals global energy use at night by the human population. Just as biological science does, it invites you to think more deeply about the world of life—and about our impact upon it.

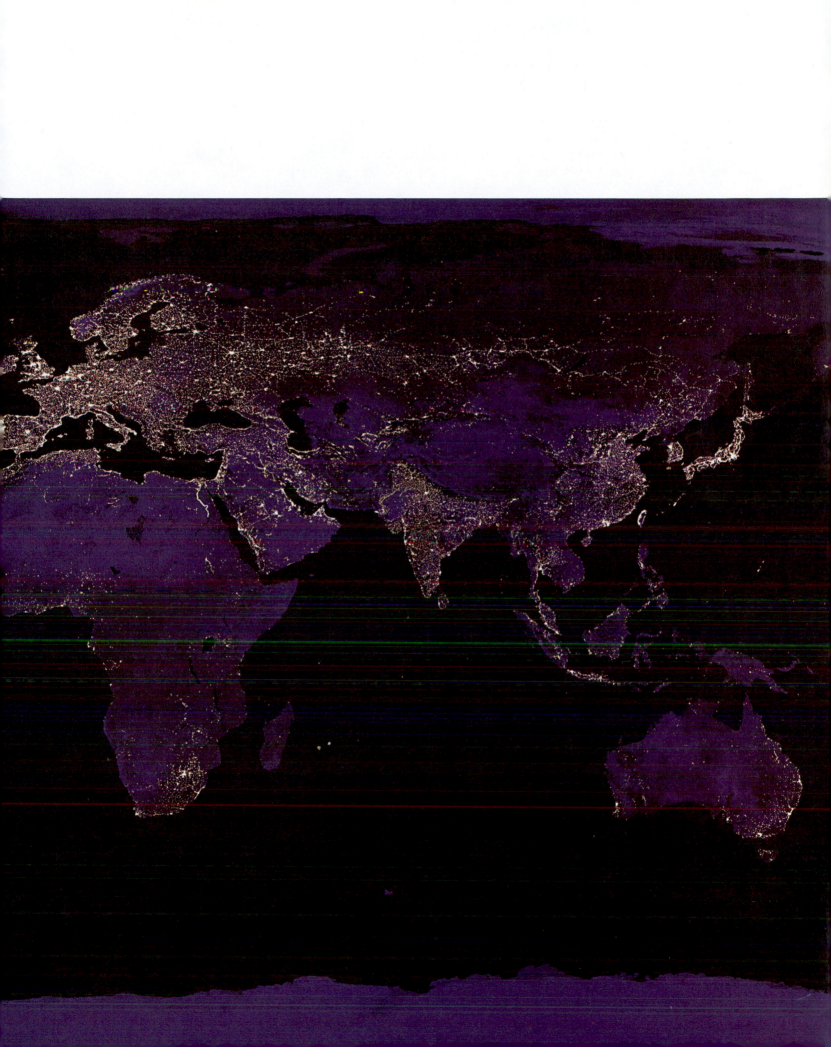

III Principles of Evolution

Two male frigate birds (*Fregata minor*) in the Galápagos Islands, far from the coast of Ecuador. Each male inflates a gular sac, a balloon of red skin at his throat, in a display that may catch the eye of a female. The males lurk together in the bushes, sacs inflated, until a female flies by. Then they wag their head back and forth and call out to her. Like other structures that males use only in courtship, the gular sac is probably an outcome of sexual selection—one of the topics you will read about in this unit.

17 EVIDENCE OF EVOLUTION

Measuring Time

How do you measure time? Is your comfort level with the past limited to your own generation? Probably you can relate to a few centuries of human events. But geologic time? Comprehending the distant past requires a huge intellectual leap from the familiar to the unknown.

Perhaps the possibility of an asteroid slamming into our planet can help you make the leap. Asteroids are rocky, metallic bodies hurtling through space. They are a few meters to 1,000 kilometers across. When our solar system's planets were forming, their gravitational force swept up most of the asteroids. At least 6,000 asteroids, including the one shown in Figure 17.1, still orbit the sun in a belt between Mars and Jupiter. Millions more frequently zip past Earth. They are hard to spot; they do not emit light. We cannot identify most of them until after they have passed close by. Some have passed too close for comfort.

We have evidence that big asteroid impacts influenced the history of life. For instance, researchers found a thin layer of iridium around the world, and it dates precisely to a mass extinction that wiped out the last of the dinosaurs (Figures 17.1 and 17.2). Iridium is rare on Earth but not in asteroids. There are plenty of fossils of dinosaurs below this layer. Above it, there are none, anywhere.

It has only been about 100,000 years since the first modern humans (*Homo sapiens*) evolved. We have fossils of dozens of humanlike species that lived in Africa during the 5 million years before our own species even showed up. So why are we the only ones left?

Unlike today's large, globally dispersed populations of humans, those early species lived in small bands. What if most were casualties of the twenty asteroids that collided with Earth while they were alive? What if *our* ancestors were just plain lucky? About 2.3 million years ago, a huge object from space hit the ocean west of what is now Chile. If it had struck the rotating Earth just a few hours earlier, it would have hit southern Africa instead of the ocean. Our ancestors could have been incinerated.

Now that we know what to look for, we are seeing more and more craters in satellite images of Earth. One crater, in Iraq, is less than 4,000 years old. Energy released during that particular impact was equivalent to the detonation of hundreds of nuclear weapons.

Watch the video online!

Figure 17.1   *Left,* an asteroid nineteen kilometers (about twelve miles) long, hurtling through space. *Right,* part of the worldwide, iridium-rich layer of sediment (*black*) that dates to the Cretaceous–Tertiary (K–T) boundary. This vertical section through stacked layers of rock is evidence of an asteroid impact. The red pocketknife gives you an idea of its thickness.

IMPACTS, ISSUES

Figure 17.2 Artist's interpretation and computer-generated model for the last few minutes of the Cretaceous.

If we can figure out what an asteroid impact will do to us, then we can figure out how impacts affected life in the past. We *can* comprehend life long before our own, including its brushes with good and bad cosmic luck.

You are about to make an intellectual leap through time, to places that were not even known about a few centuries ago. We invite you to launch yourself from this premise: *Any aspect of the natural world, past as well as present, has one or more underlying causes.*

That premise is the foundation for scientific research into the history of life. It guides probes into physical and chemical aspects of Earth. It guides the studies of fossils and comparisons of species. It guides tests of hypotheses by way of experiments, models, and new technologies. This research represents a shift from experience to inference— from the known to what can only be surmised. And it has given us astonishing glimpses into the past.

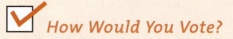

How Would You Vote?

A large asteroid could obliterate civilization and much of Earth's biodiversity. Should nations around the world contribute resources to locating and tracking asteroids? See BiologyNow for details, then vote online.

Key Concepts

EMERGENCE OF EVOLUTIONARY THOUGHT

As long ago as the 1400s, Western scientists started to learn about previously unknown species and how organisms are distributed around the world. They documented similarities and differences in the traits among living organisms and among fossil species that were being unearthed in layers of sedimentary rock. Sections 17.1, 17.2

A THEORY TAKES FORM

The emerging discoveries suggested that evolution, or changes in lines of descent, had occurred. Charles Darwin and Alfred Wallace independently developed a theory of natural selection to explain how the heritable traits that define each species might evolve. Section 17.3

EVIDENCE FROM FOSSILS

The fossil record will never be complete, so gaps are to be expected. Even so, it offers powerful evidence of change in lines of descent over great time spans. Sections 17.4, 17.5

EVIDENCE FROM BIOGEOGRAPHY

Evolutionary theories, reinforced by plate tectonics theory, help explain patterns in the distribution of species through the environment and through time. Section 17.6

EVIDENCE FROM COMPARATIVE MORPHOLOGY

The adults and embryos of different lineages often show similarities in one or more body parts that hint at descent from a common ancestor. Sections 17.7, 17.8

EVIDENCE FROM COMPARATIVE BIOCHEMISTRY

Today, researchers are using biochemical and molecular comparisons to illuminate the history of life and to clarify the evolutionary relationships among diverse species and lineages. Section 17.9

Links to Earlier Concepts

Section 1.4 sketched out key premises of the theory of natural selection. Here you will read about evidence that led to its formulation. As you read, remember that science does not deal with the supernatural, and it cannot answer subjective questions about nature (1.7).

You may wish to refresh your memory of radioisotopes (2.2), protein structure (3.5), mutation (1.4, 3.6, 14.4), nucleic acid hybridization (16.2), and automated gene sequencing (16.3). These topics are basic to understanding the biochemical and molecular comparisons that are clarifying evolutionary relationships among species.

17.1 Early Beliefs, Confounding Discoveries

LINK TO
SECTION
1.7

Prevailing beliefs can influence how we interpret clues to natural processes and their observable outcomes.

QUESTIONS FROM BIOGEOGRAPHY

Two thousand years ago the seeds of biological inquiry were taking hold in the West. Aristotle was foremost among the early naturalists. There were no books or instruments to guide him, and yet he was more than a collector of random observations. In his descriptions we see evidence that he was connecting observations in an attempt to explain the order of things. As others did, he saw nature as a continuum of organization, from lifeless matter through complex forms of plants and animals. By the fourteenth century, scholars had transformed his ideas into a rigid view of life. A Chain of Being was seen as extending from "lowest" forms to humans, and on to spiritual beings. Each kind of being, or **species**, was one separate link in the chain. All links had been designed and forged at the same time at one center of creation. They had not changed since. Once naturalists discovered and described all the links, the meaning of life would be revealed.

Then Europeans embarked on their globe-spanning explorations. They soon discovered that the world is a lot bigger than Europe. Tens of thousands of unique plants and animals from Asia, Africa, the New World, and the Pacific islands were brought back home to be carefully catalogued as one more link in the chain.

Later, Alfred Wallace and a few other naturalists moved beyond cataloguing species for its own sake. They started to identify *patterns* in where species live and how species might or might not be related. They were pioneers in **biogeography**: the study of patterns in the geographic distribution of individual species and entire communities. They were the first to think about the ecological and evolutionary forces in play.

Some patterns were intriguing. For example, many plants and animals are found only on islands in the middle of the ocean and other remote places. Many species that are strikingly similar live far apart, and vast expanses of open ocean or impassable mountain ranges keep them separated, in isolation.

Consider: Flightless, long-necked, long-legged birds are native to three continents (Figure 17.3a–c). Why are they so much alike? The plants in Figure 17.3d,e live on separate continents. Both have spines, tiny leaves, and short fleshy stems. Why are *they* so much alike?

Curiously, the flightless birds all live in the same kind of environment about the same distance from the equator. They sprint about in flat, open grasslands of dry climates, and they raise their long necks to keep an eye on predators in the distance. Both plant species also live the same distance from the equator in the

Figure 17.3 Species that resemble one another, strikingly so, even though they are native to distant geographic realms.

(a) South American rhea, (b) Australian emu, and (c) African ostrich. All three types of birds live in similar habitats. They are unlike most birds in several traits, most notably in their long, muscularized legs and their inability to get airborne.

(d) A spiny cactus native to the hot deserts of the American Southwest. (e) A spiny spurge native to southwestern Africa.

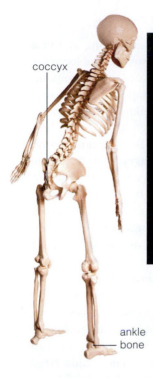
coccyx

ankle
bone

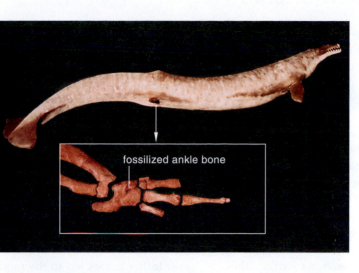
fossilized ankle bone

Figure 17.4 Body parts that have no apparent function. *Above*, reconstruction of an ancient whale (*Basilosaurus*), with a head as long as a sofa. This marine predator was fully aquatic, so it did not use its hindlimbs to support body weight as you do—yet it had ankle bones. We use our ankles, but not our coccyx bones.

Figure 17.5 Fossilized ammonites. These large marine predators lived hundreds of millions of years ago. Their shell is similar to the shell of a modern chambered nautilus.

same kind of environment—hot deserts—where water is seasonally scarce. Both species store water in their fleshy stems. Their stems have a notably thick, water-conserving cuticle and rows of sharp spines that deter thirsty, hungry herbivores.

If all birds and plants were created in one place, then how did such similar kinds end up in the same kind of environment in such distant, remote places?

QUESTIONS FROM COMPARATIVE MORPHOLOGY

The similarities and differences among species raised questions that gave rise to **comparative morphology**: the study of body plans and structures among groups of organisms. For instance, the bones in a human arm, whale flipper, and bat wing differ in size, shape, and function. As Section 17.7 explains, the different bones are located in similar body regions. They are made of the same kinds of tissues that are arranged in similar patterns. They also develop in much the same way in embryos. Naturalists who discovered the similarities wondered: Why do species that differ so much in some features look so much alike in other features?

By one hypothesis, body plans are so perfect there was no need to make a new design for each organism at the time of its creation. Yet if that were so, then why did some organisms have useless body parts? For instance, an ancient aquatic whale had ankle bones but it did not walk (Figure 17.4). Why the bones? Our coccyx is like some tailbones in many other mammals. We do not have a tail. Why do we have parts of one?

QUESTIONS ABOUT FOSSILS

About the same time, geologists were mapping layers of rock exposed by erosion or quarrying. As you will read later on, they added to the confusion when they found fossils in the same kinds of layers in different parts of the world. **Fossils** came to be recognized as the stone-hard evidence of earlier forms of life. Figures 17.4 and 17.5 have examples.

A puzzle: Many deep layers held fossils of simple marine life. Some layers above them contained fossils that were structurally similar but more intricate. In higher layers, fossils were like modern species. What did sequences in complexity among fossils of a given type mean? Were they evidence of lines of descent?

Taken as a whole, the findings from biogeography, comparative morphology, and geology did not fit with prevailing beliefs of the nineteenth century. Scholars floated novel hypotheses. If a simultaneous dispersal of all species from a center of creation was unlikely, *then perhaps species originated in more than one place*. If species had not been created in a perfect state—and fossil sequences and "useless" body parts implied they had not—*then perhaps species had become modified over time*. Awareness of evolution was in the wind.

Awareness of biological evolution emerged over centuries, through the cumulative observations of many naturalists, biogeographers, comparative anatomists, and geologists.

17.2 A Flurry of New Theories

LINK TO
SECTION
1.7

Nineteenth-century naturalists found themselves trying to reconcile the evidence of change with a traditional conceptual framework that simply did not allow for it.

SQUEEZING NEW EVIDENCE INTO OLD BELIEFS

A respected anatomist, Georges Cuvier, was among those trying to make sense of the growing evidence for change. For years he had compared fossils with living organisms. He was aware of the abrupt changes in the fossil record and was the first to recognize that they marked times of mass extinctions.

By Cuvier's hypothesis, a single time of creation had populated the world, which was an unchanging stage for the human drama. Monstrous earthquakes, floods, and other major catastrophes did happen, and many people died. Each time, survivors repopulated the world. By Cuvier's reckoning, there were no *new* species. Naturalists simply had not yet found all of the fossils that would date to the time of creation.

His hypothesis enjoyed support for a long time. It even became elevated to the rank of theory, one that later became known as **catastrophism**.

Still, many other scholars kept at the puzzle. For example, in Jean Baptiste Lamarck's view, offspring inherit traits that a parent *acquired in its lifetime*. By his hypothesis, environmental pressure and internal needs promote permanent changes in an individual's body form and functions, which offspring then inherit. By this proposed process, life was created long ago in a simple state, and it gradually improved. The force for change was an intense drive toward perfection, up the Chain of Being. Lamarck thought that the force, centered in nerves, directed an unknown "fluida" to body parts needing change.

Try using his hypothesis to explain a giraffe's long neck. Suppose a short-necked, hungry ancestor of the modern giraffe kept stretching its neck in order to browse on leaves beyond the reach of other animals. Lengthier and lengthier stretches directed fluida into its neck and thus made the neck permanently longer. Offspring inherited a longer neck, and they stretched their necks, too. Generations that strained to reach ever loftier leaves led to the modern giraffe.

As Lamarck correctly inferred, the environment *is* a factor in changes in lines of descent. However, his hypothesis, and Cuvier's, has not been supported by experimental tests. Environmental factors can alter an individual's phenotype, as when a male builds large muscles through strength training. But any child of a muscle-bound parent will not be born muscle-bound. It can inherit genes, but not increased muscle mass.

VOYAGE OF THE *BEAGLE*

In 1831, in the midst of the confusion, Charles Darwin was twenty-two years old and wondering what to do with his life. Ever since he was eight, he had wanted

Figure 17.6 (**a**) Charles Darwin. (**b**) Replica of the *Beagle* sailing off a rugged coastline of South America. During one of his trips, Darwin ventured into the Andes. He discovered fossils of marine organisms in rock layers 3.6 kilometers above sea level.

(**c–e**) The Galápagos Islands are isolated in the ocean, far to the west of Ecuador. They arose by volcanic action on the seafloor about 5 million years ago. Winds and currents carried organisms to the once-lifeless islands. All of the native species are descended from those travelers. At far right, a blue-footed booby, one of many species Darwin observed during his voyage.

to hunt, fish, collect shells, or just watch insects and birds—anything but sit in school. Later, at his father's insistence, he did attempt to study medicine in college. The crude, painful procedures being used on patients in that era sickened him. His exasperated father urged him to become a clergyman, and so Darwin packed for Cambridge. His grades were good enough to earn a degree in theology. Yet he spent most of his time with faculty members who embraced natural history.

John Henslow, a botanist, perceived Darwin's real interests. He hastily arranged for Darwin to become ship's naturalist aboard the *Beagle*, which was about to leave on a five-year voyage around the world. The young man who had hated school and had no formal training quickly became an enthusiastic naturalist.

The *Beagle* sailed first to South America to finish work on mapping the coastline (Figure 17.6). During the Atlantic crossing, Darwin collected and studied marine life. He read Henslow's parting gift, the first volume of Charles Lyell's *Principles of Geology*. He saw diverse species in environments ranging from sandy shores of remote islands to high mountains. He also started circling the question of evolving life, which was now on the minds of many individuals.

Darwin started by mulling over a radical theory. As Lyell and other geologists were arguing, erosion and other gradual, natural processes of change had more impact on Earth history than rare catastrophes. Geologists for years had chipped away at sandstones,

limestones, and other rocks that form after sediments slowly accumulate in the beds of lakes, rivers, and seas. They took earthquakes and other less frequent events into account. As they knew, immense floods, more than a hundred big earthquakes, and twenty or so volcanic eruptions happen in a typical year, which means catastrophes are not that unusual.

The idea that gradual, repetitive change had shaped Earth became known as the **theory of uniformity**. It challenged the prevailing views of Earth's age.

The theory bothered scholars who firmly believed that Earth could be no more than 6,000 years old. They believed people had recorded all that had happened in those 6,000 years—and in all that time, no one had mentioned seeing a species evolve. Even so, by Lyell's calculations, it must have taken millions of years to sculpt the present landscape. *Was that not enough time for species to evolve in many diverse ways?* Later, Darwin thought so. But exactly *how* did they evolve? He would end up devoting the rest of his life to that burning question.

Prevailing beliefs can influence how we interpret clues to natural processes and their observable outcomes.

Darwin's observations during a global voyage helped him think about species in a novel way.

route of Beagle

Darwin
Wolf
EQUATOR
Galápagos Islands
Pinta
Marchena   Genovesa
Santiago   EQUATOR
Bartolomé
Seymour
Rábida Baltra
Fernandina   Pinzón
Santa Cruz
Santa Fe
San Cristóbal
Tortuga
Isabela
Española
Floreana

17.3 Darwin, Wallace, and Natural Selection

LINK TO
SECTION
1.4

Darwin's observations of thousands of species in different parts of the world helped him see how species might evolve.

OLD BONES AND ARMADILLOS

Darwin brought thousands of specimens with him to England. Although he had page after page of notes, he had been careless about recording where each species lived and what its habitat was like. He had left much of the "geography" out of biogeography. Colleagues helped him fill in some of the blanks, and in time he was able to explain how species might evolve.

Among the specimens were fossils of glyptodonts from Argentina. Of all animals, only living armadillos are like the now-extinct glyptodonts (Figure 17.7). Of all places on Earth, armadillos live only in the places where glyptodonts once lived.

If these animals had been created at the same time, lived in the same place, and were so alike in certain odd traits—such as body armor made of overlapping scales—then why is only one still with us? What if the glyptodonts were ancient relatives of armadillos? What if some traits of their common ancestor had changed in the line of descent that led to armadillos? *Descent with modification*—it did seem possible. What, then, could be the driving force for evolution?

A KEY INSIGHT—VARIATION IN TRAITS

While Darwin assessed his notes, an essay by Thomas Malthus, a clergyman and economist, made him reflect on a topic of social interest. Malthus had correlated population size with famine, disease, and war. He said that humans run out of food, living space, and other resources because they reproduce too much. The larger a population gets, the more individuals there are to reproduce. Individuals compete with one another for dwindling resources. Many starve, get sick, or engage in war or other forms of competition.

Darwin deduced that *any* population has a capacity to produce more individuals than the environment can support. Even one sea star can produce 2,500,000 eggs per year, but the seas do not fill with sea stars. (For one thing, predators eat many of the eggs and larvae.)

Darwin also reflected on species he had observed during his voyage. Individuals of those species were not alike in their details. They varied in size, color, and other traits. *It dawned on Darwin that variations in traits influence an individual's ability to secure resources and to survive and reproduce in the environment.*

He thought about the Galápagos Islands, separated from South America by 900 kilometers of open ocean. Nearly all of their finch species live nowhere else, yet they share traits with mainland species. Perhaps fierce storms had blown a few mainland birds out to sea. Perhaps prevailing winds and currents had dispersed them to the Galápagos. Perhaps those modern species were island-hopping descendants of the colonizers.

As he knew, different species live in diverse habitats near coasts, in dry lowlands, and in mountain forests. One strong-billed type is better than others at cracking open hard seeds (Figure 17.8). A drought that lasts for several years will make soft seeds harder to find. A strong-billed individual will survive and reproduce more than the others. Its bill size has a heritable basis, so its offspring will be favored, also. Conditions in the prevailing environment "select" individuals that have strong bills, which in time become more frequent in the population. *And a population is evolving when forms of heritable traits change over the generations.*

Figure 17.7 (**a**) From Texas, a modern armadillo, about a foot long excluding the tail. (**b**) A Pleistocene glyptodont, which was about as big as a Volkswagen Beetle, and now extinct. Glyptodonts shared unusual traits and a restricted distribution with the existing armadillos. Yet the two kinds of animals are widely separated in time. Their similarities were a clue that helped Darwin develop a theory of evolution by natural selection.

a

b

Figure 17.8 Three of thirteen finch species on the Galápagos Islands. (**a**) A big-billed seed cracker, *Geospiza magnirostris.* (**b**) *G. scandens* eats cactus fruit and insects in cactus flowers. (**c**) *Camarhynchus pallidus* uses cactus spines and twigs to probe for wood-boring insects. Differences in bill shape depend in large part on when and where the signaling molecule BMP4 is switched on in bird embryos. Mutations in regulatory elements may cause the bill variations. Compare the different bills of Hawaiian honeycreepers (Chapter 19).

NATURAL SELECTION DEFINED

Let's now put Darwin's observations and conclusions in the context of what we have learned from genetics and molecular biology:

1. *Observation:* Natural populations have an inherent reproductive capacity to increase in size over time.

2. *Observation:* No population can indefinitely grow in size, because its individuals will run out of food, living space, and other resources.

3. *Inference:* Sooner or later, individuals will end up competing for dwindling resources.

4. *Observation:* Individuals share a pool of heritable information about traits, encoded in genes.

5. *Observation:* Variations in traits start with alleles, slightly different molecular forms of genes that arise through mutations.

6. *Inferences:* Some forms of traits prove better than others at helping an individual compete for resources, survive, and reproduce. In time, alleles for adaptive forms become more frequent relative to other alleles in the population. They lead to increased **fitness**—an increase in adaptation to the environment as measured by the genetic contribution to future generations.

7. *Conclusions:* **Natural selection** is the outcome of differences in reproduction among individuals of a population that vary in shared traits. Environmental agents of selection act on the range of variation, and the population may evolve as a result.

Darwin kept on looking for patterns in his data and filling in gaps in his reasoning. He also wrote out his theory but let ten years pass without publishing it. He waited too long. Alfred Wallace sent him an essay he was working on that outlined the same theory! Today, Wallace is known as the father of biogeography. He did brilliant fieldwork in the Amazon River Basin, Malay Archipelago, and elsewhere (Figure 17.9). He had written earlier letters to Lyell and Darwin about patterns in the geographic distribution of species. He, too, had connected the dots.

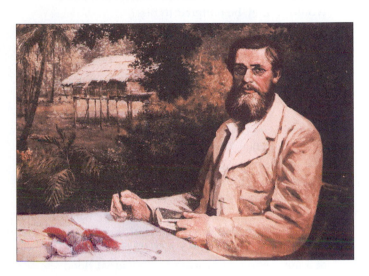

Figure 17.9 Alfred Wallace. For one account of the Darwin–Wallace story, read David Quammen's *Song of the Dodo*.

In 1858, just weeks after Darwin received Wallace's essay, their similar theories were presented jointly at a scientific meeting. Wallace was still in the field and knew nothing about the meeting, which Darwin did not attend. The next year, Darwin published *On the Origin of Species*, which laid out detailed evidence in support of his theory.

You may have heard that Darwin's book fanned an intellectual firestorm, but most scholars were quick to accept the idea that diversity is a result of evolution. The theory of natural selection *was* fiercely debated. Decades passed before experimental evidence from a new field, genetics, led to its widespread acceptance.

As Darwin and Wallace perceived, natural selection is the outcome of differences in survival and reproduction among traits. Natural selection can lead to increased adaptation to the environment, as measured by fitness—the relative genetic contribution to future generations.

17.4 Fossils—Evidence of Ancient Life

Turn now to fossil evidence of the connection between life's evolution and the evolution of Earth.

About 500 years ago, Leonardo da Vinci was puzzled by seashells entombed in the rocks of northern Italy's high mountains, hundreds of kilometers from the sea. How did they get there? By the prevailing belief, water from a stupendous and divinely invoked flood had surged up into the mountains, where it deposited the shells. But many shells were thin, fragile, and intact. If they had been swept across such great distances, then wouldn't they be battered to bits?

Leonardo also brooded about the rocks. They were stacked like cake layers. Some layers had shells, others had none. Then he remembered how large rivers swell with spring floodwaters and deposit silt in the sea. Did such depositions happen in ancient seasons? If so, then shells in the mountains could be evidence of layered communities of organisms that once lived in the seas!

By the 1700s, fossils were being accepted as remains and impressions of organisms that lived in the past. (*Fossil* comes from the Latin word for "something that was dug up.") People were still interpreting fossils through the prism of cultural beliefs, as when a Swiss naturalist unveiled the remains of a giant salamander and excitedly announced that they were the skeleton of a man who had drowned in the great flood.

By midcentury, naturalists were questioning these interpretations. Mining, quarrying, and excavations for canals were under way. Diggers were discovering similar rock layers and similar sequences of fossils in distant places, such as the cliffs on both sides of the English Channel. If those layers had been deposited over time, then a vertical series of fossils embedded in them might be a record of past life—*a fossil record*.

HOW DO FOSSILS FORM?

Most fossils are bones, teeth, shells, seeds, spores, and other hard parts (Figure 17.10). Even fossilized feces (coprolites) hold parts of organisms that were eaten. Indirectly, imprints of leaves, stems, tracks, burrows, and other *trace* fossils offer more evidence of past life.

Fossilization is a slow process that starts when an organism or traces of it become covered by sediments or volcanic ash. Water slowly infiltrates the remains, and metal ions and other inorganic compounds that are dissolved in it replace the minerals in bones and other hardened tissues. As sediments accumulate, they exert increasing pressure on the burial site. In time, the pressure and mineralization processes transform those remains into stony hardness.

Remains that become buried quickly are less likely to be obliterated by scavengers. Preservation is also favored when a burial site stays undisturbed. Usually, however, erosion and other geologic assaults deform, crush, break, or scatter the fossils. This is one reason fossils are relatively rare.

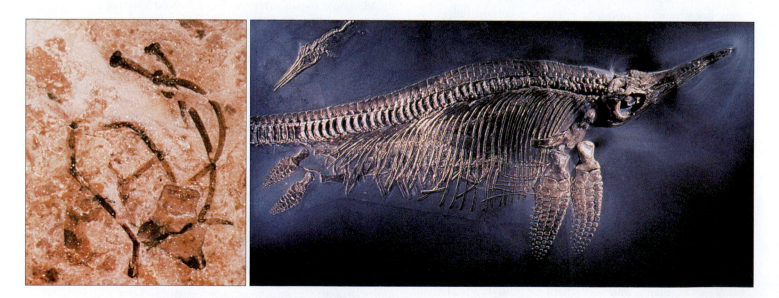

Figure 17.10 Two of the more than 250,000 species known from the fossil record. *Left*, fossilized parts of the oldest known land plant (*Cooksonia*). Its stems were a little taller than the length of a toothpick. *Right*, fossilized skeleton of an ichthyosaur. This marine reptile lived 200 million years ago.

Figure 17.11 A slice through time—Butterloch Canyon, Italy, once at the bottom of a sea. Its sedimentary rock layers slowly formed over hundreds of millions of years. Later, geologic forces lifted the stacked layers above sea level. Later still, the erosive force of river water carved the canyon walls and exposed the layers. Scientists Cindy Looy and Mark Sephton are climbing to reach the Permian–Triassic boundary layer, where they will look for fossilized fungal spores.

Other factors affect preservation. Organic materials cannot decompose in the absence of free oxygen, for instance. They may endure when sap, tar, ice, mud, or another air-excluding substance protects them. Insects in amber and frozen woolly mammoths are examples.

FOSSILS IN SEDIMENTARY ROCK LAYERS

Stratified (stacked) layers of sedimentary rock formed long ago from deposits of volcanic ash, silt, sand, and other materials. Sand and silt piled up after rivers transported them from land to the sea, as Leonardo suspected. Sandstones formed from sand, and shales from silt. Depositions were sometimes interrupted, in part because the sea level changed as ice ages began. Tremendous volumes of water froze in glaciers, rivers dried up, and the depositions ended in some regions. Later in time, when the climate warmed and glaciers melted, depositions resumed.

The formation of sedimentary rock layers is called **stratification**. The deepest layer was the first to form; those closest to the surface were the last. Most formed horizontally, as in Figure 17.11, because particles tend to settle in response to gravity. You may see tilted or ruptured layers, as along a road that was cut into a mountainside. Major crustal movements or upheavals disturbed them after they formed.

Most fossils are in sedimentary rock. Understand how rock layers form, and you know that fossils in them formed at specific times in the past. Specifically, *the older the rock layer, the older the fossils.*

INTERPRETING THE FOSSIL RECORD

We have fossils for more than 250,000 known species. Judging from the current range of biodiversity, there must have been many, many millions more. Yet the fossil record will never be complete. Why is this so?

The odds are against finding evidence of an extinct species. Why? At least one specimen had to be buried gently before it decomposed or something ate it. It had to escape erosion, flowing lava, and other forces of nature. The fossil had to end up where someone can find it. For instance, many have become exposed on canyon walls after a river or glacier slowly carved its way through sedimentary rock layers (Figure 17.11).

Also, most ancient species did not lend themselves to preservation. Unlike bony fishes and hard-shelled mollusks, for instance, the soft-bodied jellyfishes and worms do not show up as much in the fossil record. Probably they were just as common, or more so.

Also think about population density and body size. One plant population might release millions of spores in a single season. The earliest humans lived in small bands and raised few offspring. What are the odds of finding even one fossilized human bone compared to finding spores of plants that lived at the same time?

Finally, imagine one line of descent, a **lineage**, that vanished when its habitat on a remote volcanic island sank into the sea. Or imagine two lineages, one lasting only briefly and the other for billions of years. Which is more likely to be represented in the fossil record?

> Fossils are physical evidence of organisms that lived in the remote past, a stone-hard historical record of life. In general, the oldest are in the deepest sedimentary rocks.
>
> The fossil record is incomplete. Geologic events obliterated much of it. The record is slanted toward species that had hard parts, dense populations, and wide distribution, and that persisted a long time.
>
> Even so, the fossil record is now substantial enough to help us reconstruct patterns and trends in the history of life.

17.5 Dating Pieces of the Puzzle

LINK TO
SECTION
2.2

How do we assign fossils to a place in time? In other words, how do we know how old fossils really are?

RADIOMETRIC DATING

At one time, people could assign only *relative* ages to their fossil treasures, not absolute ones. For instance, a fossilized mollusk embedded in a layer of rock was said to be younger than a fossil below it and older than a fossil above it, and so forth.

Things changed with **radiometric dating**. This is a way to measure proportions of a daughter isotope and the parent radioisotope of some element trapped in a rock since the time the rock formed. A radioisotope is a form of an element with an unstable nucleus (Section 2.2). Its atoms lose energy and subatomic particles—they decay—until they reach a more stable form.

We cannot predict the exact instant of one atom's decay, but a predictable number of a radioisotope's atoms decay in a characteristic time span. Like the ticking of a perfect clock, the rate of decay for each isotope is constant. Changes in pressure, temperature, or chemical state do not alter it. The time it takes for half of a quantity of a radioisotope's atoms to decay is its **half-life** (Figure 17.12*a*).

For instance, uranium 238's half-life is 4.5 billion years. It decays into thorium 234, which then decays into something else, and so on through intermediate isotopes to lead, the final, stable daughter element for this series. By measuring the uranium 238/lead ratio in the oldest known rocks, geologists figured out that Earth formed more than 4.6 billion years ago.

Radiometric dating has an error factor of less than 10 percent. Recent fossils still hold some carbon and can be dated on the basis of their carbon 14/carbon 12 ratio, as in Figure 17.12*b–d*. Older fossils are dated on the basis of isotope ratios in volcanic rocks or ashes buried with them in the same sedimentary layer.

PLACING FOSSILS IN GEOLOGIC TIME

Early geologists carefully counted backward through layers of sedimentary rock, then used their counts to construct a chronology of Earth history, or a **geologic time scale** (Figure 17.13). By comparing evidence from around the world, they found four abrupt transitions in fossil sequences and used them as boundaries for four great intervals. They named the first interval the Proterozoic, to indicate that it predates fossils of early animals. They named different intervals the Paleozoic,

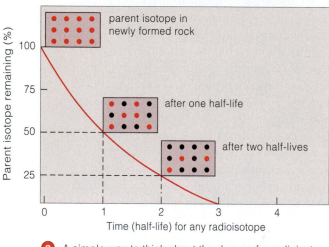

a A simple way to think about the decay of a radioisotope to a more stable form, as plotted against time.

Figure 17.12 *Animated!* (**a**) The decay of radioisotopes at a fixed rate to more stable forms. The half-life of each kind of radioisotope is the time it takes for 50 percent of a given sample to decay. After two half-lives, 75 percent of the sample has decayed, and so on.

(**b–d**) Radiometric dating of a fossil. Carbon 14 (^{14}C) forms in the atmosphere. There, it combines with free oxygen, the result being carbon dioxide. Along with far greater quantities of its more stable isotopes, trace amounts of carbon 14 enter food webs by way of photosynthesis. All organisms incorporate carbon into body tissues.

b Long ago, trace amounts of ^{14}C and a lot more ^{12}C were incorporated into the tissues of a living mollusk. Those atoms of carbon were part of organic compounds making up tissues of its prey. As long as that mollusk lived, the proportion of ^{14}C to ^{12}C in its tissues remained the same.

c When the mollusk died, it stopped gaining carbon. Over time, the proportion of ^{14}C to ^{12}C in its remains fell because of the radioactive decay of ^{14}C. Half of the ^{14}C had decayed in 5,370 years, half of what was left was gone after another 5,370 years, and so on.

d Fossil hunters find the fossil. They measure its ^{14}C/^{12}C ratio to determine half-life reductions since death. In this example, the ratio turns out to be one-eighth of the ^{14}C/^{12}C ratio in living organisms. Thus the mollusk lived about 16,000 years ago.

Eon	Era	Period	Epoch	Millions of Years Ago	Major Geologic and Biological Events That Occurred Millions of Years Ago (mya)
PHANEROZOIC	CENOZOIC	QUATERNARY	Recent	0.01	1.8 mya to present. Major glaciations. Modern humans evolve. The most recent *extinction crisis* is under way.
			Pleistocene	1.8	
		TERTIARY	Pliocene	5.3	65–1.8 mya. Major crustal movements, collisions, mountain building. Tropics, subtropics extend poleward. When climate cools, dry woodlands, grasslands emerge. *Adaptive radiations* of flowering plants, insects, birds, mammals.
			Miocene	22.8	
			Oligocene	33.7	
			Eocene	55.5	
			Paleocene	65	65 mya. Asteroid impact; *mass extinction* of all dinosaurs and many marine organisms.
	MESOZOIC	CRETACEOUS	Late	99	99–65 mya. Pangea breakup continues, inland seas form. Adaptive radiations of marine invertebrates, fishes, insects, and dinosaurs. Origin of angiosperms (flowering plants).
			Early	145	145–99 mya. Pangea starts to break up. Marine communities flourish. *Adaptive radiations* of dinosaurs.
		JURASSIC		213	145 mya. Asteroid impact? Mass extinction of many species in seas, some on land. Mammals, some dinosaurs survive.
		TRIASSIC		248	248–213 mya. *Adaptive radiations* of marine invertebrates, fishes, dinosaurs. Gymnosperms dominate land plants. Origin of mammals.
	PALEOZOIC	PERMIAN		286	248 mya. *Mass extinction.* Ninety percent of all known families lost.
					286–248 mya. Supercontinent Pangea and world ocean form. On land, *adaptive radiations* of reptiles and gymnosperms.
		CARBONIFEROUS		360	360–286 mya. Recurring ice ages. On land, *adaptive radiations* of insects, amphibians. Spore-bearing plants dominate; cone-bearing gymnosperms present. Origin of reptiles.
		DEVONIAN		410	360 mya. *Mass extinction* of many marine invertebrates, most fishes.
					410–360 mya. Major crustal movements. Ice ages. *Mass extinction* of many marine species. Vast swamps form. Origin of vascular plants. *Adaptive radiation* of fishes continues. Origin of amphibians.
		SILURIAN		440	440–410 mya. Major crustal movements. *Adaptive radiations* of marine invertebrates, early fishes.
		ORDOVICIAN		505	505–440 mya. All land masses near equator. Simple marine communities flourish until origin of animals with hard parts.
		CAMBRIAN		544	544–505 mya. Supercontinent breaks up. Ice age. *Mass extinction.*
PROTEROZOIC				2,500	2,500–544 mya. Oxygen accumulates in atmosphere. Origin of aerobic metabolism. Origin of eukaryotic cells. Divergences lead to eukaryotic cells, then protists, fungi, plants, animals.
ARCHEAN AND EARLIER					3,800–2,500 mya. Origin of photosynthetic prokaryotic cells. 4,600–3,800 mya. Origin of Earth's crust, first atmosphere, first seas. Chemical, molecular evolution leads to origin of life (from proto-cells to anaerobic prokaryotic cells).

Figure 17.13 *Animated!*   Geologic time scale. *Red* boundaries mark times of the greatest mass extinctions. If these spans were to the same scale, the Archean and Proterozoic portions would extend downward, spill off the page, and spill across the room. Compare Figure 17.14.

Mesozoic, and lastly the "modern" era, the Cenozoic. Researchers now correlate the geologic time scale with **macroevolution**, or major patterns, trends, and rates of change among lineages. Also, they have subdivided the Proterozoic into finer intervals because it was far more immense than early researchers suspected. Life originated in one of those intervals, the Archean eon.

The geologic time scale now has absolute dates assigned to its boundaries, based on radiometric dating methods. The time scale has been correlated with macroevolution: major patterns, trends, and rates of change among lineages.

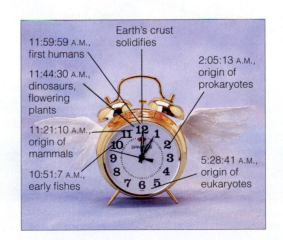

Earth's crust solidifies
11:59:59 A.M., first humans
2:05:13 A.M., origin of prokaryotes
11:44:30 A.M., dinosaurs, flowering plants
11:21:10 A.M., origin of mammals
10:51:7 A.M., early fishes
5:28:41 A.M., origin of eukaryotes

Figure 17.14   How time flies—a geologic time clock. Think of the spans as minutes on a clock that runs from midnight to noon. The recent epoch started after the very last 0.1 second before noon. And where does that put you?

17.6 Drifting Continents, Changing Seas

By clinking their hammers against the rocks, early geologists realized that "solid earth" does not stay put. It moves.

When geologists were first starting to map the vertical stacks of sedimentary rock, the theory of uniformity prevailed. They thought that mountain building and erosion had repeatedly altered Earth's surface in the same ways over time. Eventually, however, it became clear that those recurring geologic events were only part of the picture. Like life, the "unchanging" Earth had changed irreversibly.

AN OUTRAGEOUS HYPOTHESIS

For instance, the Atlantic coasts of South America and Africa seemed to "fit" like jigsaw puzzle pieces. Were all continents once part of a bigger one that had split into fragments and drifted apart? One model for the proposed supercontinent—**Pangea**—took into account the world distribution of fossils and existing species. It also took into account glacial deposits, which held clues to ancient climate zones.

Most scientists did not accept the continental drift hypothesis. Continents drifting on their own across Earth's mantle seemed to be an outrageous idea, and they preferred to think that continents did not move.

Evidence kept piling up. For example, iron-rich rocks are molten when they form, and their bits of iron become oriented toward Earth's magnetic poles. They stay that way after the rocks harden. Yet in North and South America, the tiny iron compasses in rocks that formed 200 million years ago didn't point to the north and south poles. So scientists made a map to fit the north–south alignment of their compasses. Their map put North America and western Europe right next to each other, no Atlantic Ocean between them.

Later, deep-sea probes revealed that the seafloor is spreading away from the mid-oceanic ridges (Figure 17.15). Molten rock that is spewing from a ridge flows sideways in both directions, then it hardens into new crust. The formation of more crust forces older crust into trenches elsewhere in the seafloor. These ridges and trenches are the edges of enormous crustal plates, like pieces of a gargantuan cracked eggshell. They all move at almost imperceptibly slow rates. Over great time spans, land masses end up in different locations.

These findings put continental drift into a broader explanation of crustal movements, now known as the

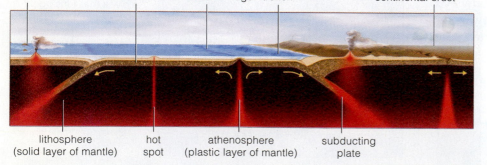

a

b

island arc   oceanic crust   oceanic ridge   trench   continental crust

lithosphere (solid layer of mantle)   hot spot   athenosphere (plastic layer of mantle)   subducting plate

Figure 17.15 Forces of geologic change.

(a) Present configuration of Earth's crustal plates. These immense, rigid portions of the crust split, drift apart, and collide at almost imperceptible rates. In Appendix VIII, this map is greatly enlarged to show details.

(b) Huge plumes of molten material drive the movement. They well up from the interior, then spread laterally under the crust and rupture it at deep, mid-oceanic ridges. The molten material seeps out, cools, and slowly hardens into new seafloor, which displaces plates away from the ridges.

The advancing edge of one plate can plow under an adjacent plate and lift it up. The Cascades, Andes, and other great mountain ranges paralleling the coasts of continents formed this way. When 2004 drew to a close, the Indian Plate lurched violently under the Eurasian Plate and caused huge tsunamis. These earthquake-generated ocean waves traveled 600 miles per hour across the Indian Ocean and killed more than 240,000 people.

Besides these forces, superplumes ruptured the crust at what are now called "hot spots" in the mantle. The Hawaiian Archipelago has been forming this way. Continents also can rupture in their interior. Deep rifting and splitting are happening now in Missouri, at Lake Baikal in Russia, and in eastern Africa.

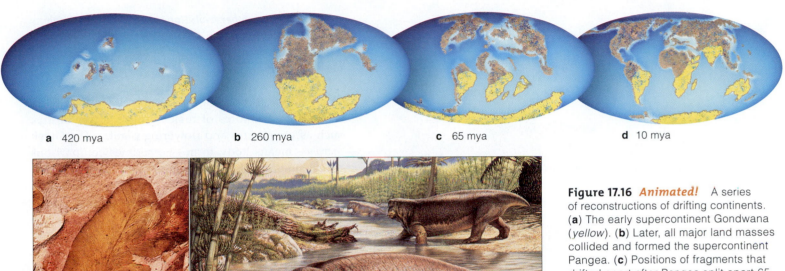

a 420 mya **b** 260 mya **c** 65 mya **d** 10 mya

Figure 17.16 *Animated!* A series of reconstructions of drifting continents. (**a**) The early supercontinent Gondwana (*yellow*). (**b**) Later, all major land masses collided and formed the supercontinent Pangea. (**c**) Positions of fragments that drifted apart after Pangea split apart 65 million years ago, and (**d**) their positions 10 million years ago.

About 260 million years ago, seed ferns and other plants lived nowhere except on the area of Pangea that had once been Gondwana. So did mammal-like reptiles named therapsids. (**e**) Fossilized leaf of one of the seed ferns, *Glossopteris*. (**f**) *Lystrosaurus*, a therapsid about 1 meter (3 feet) long. This tusked herbivore fed on fibrous plants in dry floodplains.

plate tectonics theory. Researchers soon found ways to apply the new theory's predictive power.

For example, the same series of basalt formations, coal seams, and glacial deposits occurs in Africa, India, Australia, and South America. Each of these southern continents has fossils of the seed fern *Glossopteris* and of a therapsid, *Lystrosaurus* (Figure 17.16). This plant's seeds and the therapsid were too heavy to float across the ocean from one continent to the other. Researchers suspected that the organisms had evolved together on **Gondwana**, a supercontinent that preceded Pangea.

Like other land masses in the Southern Hemisphere, Antarctica formed after Gondwana broke up. Someone predicted that fossils of *Glossopteris* and *Lystrosaurus* would be discovered in a series of basalt formations, coal seams, and glacial deposits in Antarctica. In time, explorers did find the series and the fossils. Evidence supported the prediction and plate tectonics theory.

A BIG CONNECTION

Let's take stock. In the remote past, slow movements of Earth's crustal plates put immense land masses on collision courses. Over time, land masses converged and formed supercontinents, which later split at deep rifts and formed new ocean basins. Gondwana drifted south from the tropics, across the south pole, then north until it piled into other land masses. The result was Pangea, a supercontinent that extended from pole to pole with a single world ocean lapping against its coasts. All the while, erosive forces of water and wind resculpted the land surface. Asteroids and meteorites slammed into Earth's crust. The major impacts and their aftermath caused long-term changes in the global temperature, atmosphere, and regional climates.

Such changes on land and in the ocean and atmosphere influenced life's evolution. Imagine early life in shallow, warm waters along continents. Shorelines vanished as continents collided and wiped out many lineages. Yet, even as old habitats vanished, new ones opened up for survivors—and evolution took off in new directions.

Over the past 3.8 billion years, slow movements in Earth's crust as well as catastrophic events have changed the land, the atmosphere, and the ocean. Those changes have had profound effects on the evolution of life.

17.7 Divergences From a Shared Ancestor

To biologists, remember, evolution simply means heritable changes in lines of descent. Comparisons of the body form and structures of major groups of organisms yield clues to evolutionary trends.

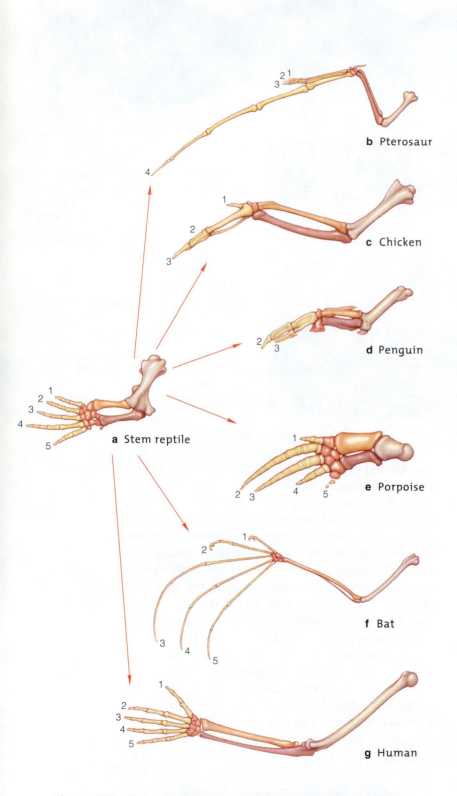

a Stem reptile

b Pterosaur

c Chicken

d Penguin

e Porpoise

f Bat

g Human

Figure 17.17 Morphological divergence among vertebrate forelimbs, starting with bones of a stem reptile (a cotylosaur). Similarities in the number and position of skeletal elements were preserved when diverse forms evolved. Some bones were lost over time (compare the numbers 1 through 5). The drawings are not to the same scale.

Comparative morphology, again, is the study of body forms and structures of major groups of organisms, such as vertebrates and flowering plants. (The Greek *morpho–* means body form.) Sometimes comparisons reveal similarities in one or more body parts between groups, which may be evidence of a common ancestor. Such body parts are **homologous structures** (*homo–* means the same). Even when different groups use the parts for different functions, the genes for constructing those parts point to shared ancestry.

MORPHOLOGICAL DIVERGENCE

Populations of a species diverge genetically after gene flow ends between them (Chapter 18). In time, some morphological traits that help to define their species commonly diverge, also. Change from the body form of a common ancestor is a major macroevolutionary pattern called **morphological divergence**.

Even if the same body part of two related species became dramatically different, some other aspects of the species may remain alike. A careful look beyond unique modifications may reveal the shared heritage.

For instance, all vertebrates on land descended from the first amphibians. Divergences led to what we call reptiles, then to birds and mammals. We know about "stem reptiles" that probably were ancestral to these groups. Fossilized, five-toed limb bones tell us that the ancestral species crouched low to the ground (Figure 17.17a). Later, descendants diversified into many new habitats on land. A few descendants that had become adapted for walking on land even returned to the seas after environmental conditions changed.

A five-toed limb was evolutionary clay. It became molded into different kinds of limbs having different functions. In lineages that eventually led to penguins and porpoises, it became modified into flippers used in swimming. In the lineage leading to modern horses, it became modified into long, one-toed limbs suitable for running fast. Among moles, it became stubby and useful for burrowing into dirt. Among elephants, it became strong and pillarlike, suitable for supporting a great deal of weight.

The five-toed limb also became modified into the human arm and hand. Later on, a thumb evolved in opposition to the four fingers of the hand. It was the basis of stronger and more precise motions.

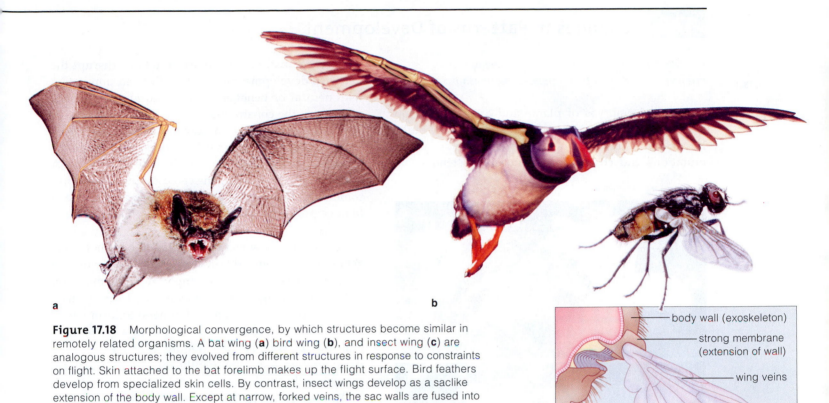

a

b

body wall (exoskeleton)

strong membrane
(extension of wall)

wing veins

c

Figure 17.18 Morphological convergence, by which structures become similar in remotely related organisms. A bat wing (**a**) bird wing (**b**), and insect wing (**c**) are analogous structures; they evolved from different structures in response to constraints on flight. Skin attached to the bat forelimb makes up the flight surface. Bird feathers develop from specialized skin cells. By contrast, insect wings develop as a saclike extension of the body wall. Except at narrow, forked veins, the sac walls are fused into a thin membrane. Wing veins, reinforced with chitin, hold airways and nerves.

Even though forelimbs are not the same in size, shape, or function from one group of vertebrates to the next, they clearly are alike in the structure and positioning of their bony elements. They also are alike in the patterns of nerves, blood vessels, and muscles that develop inside them. In addition, comparisons of the early embryos of different vertebrates reveal strong resemblances in patterns of bone development. Such similarities point to a shared ancestor.

MORPHOLOGICAL CONVERGENCE

Body parts with similar form or function in different lineages are not *always* homologous. Sometimes they evolved independently in remote lineages. Parts that differed at first may have evolved in similar ways as organisms became subjected to similar environmental pressures. **Morphological convergence** refers to cases where dissimilar body parts evolved in similar ways in evolutionarily distant lineages.

For instance, you just read about the homologous forelimbs of birds and bats. Bones aside, are bird and bat wings homologous, too? No. The flight surface of birds evolved as a sweep of feathers, all derived from skin. The forelimb structurally supports it. The flight surface for bats is a thin membrane, an extension of the skin itself. The bat wing is attached to reinforcing bony elements inside the forelimb (Figure 17.18a,b).

The insect wing, too, resembles bird and bat wings in its function—flight. Is it homologous with them? No. This wing develops as an extension of an outer body wall reinforced with chitin. No underlying bony elements support it (Figure 17.18c).

The differences between bat, bird, and insect wings are evidence that each of these animal groups adapted independently to the same physical constraints that govern how a wing can function in the environment. The wings of all three are **analogous structures**. They are not modifications of comparable body parts in different lineages. They are three different responses of different body parts to similar challenges. The Greek *analogos* means similar to one another.

With morphological divergence, comparable body parts became modified in different ways in different lines of descent from a common ancestor.

Such divergences resulted in homologous structures. Even if these body parts differ in size, shape, or function, they have an underlying similarity because of shared ancestry.

With morphological convergence, dissimilar body parts became similar in lineages that are not closely related. Such body parts are analogous structures. They converged in form only as an outcome of similar pressures.

17.8 Changes in Patterns of Development

LINKS TO
SECTIONS
14.5, 15.2, 15.3

Comparing the patterns of embryonic development often yields evidence of evolutionary relationships.

Multicelled embryos of plants and animals develop step-by-step from a fertilized egg, and there are built-in constraints on how the body plan develops. Most mutations and changes in chromosomes tend to be selected against, because most mutations disrupt the inherited developmental program. Even so, a mutation with neutral or beneficial effects can move a lineage past one of the constraints.

Master gene mutations can do this. Recall, from Sections 15.2 and 15.3, that homeotic genes guide the formation of tissues and organs in orderly patterns. A mutation in a homeotic gene can disrupt the patterns, sometimes drastically. Such disruptions typically lead to huge problems, but once in a while an altered body plan proves to be advantageous.

You already saw some examples of how homeotic genes guide when and how flowers form. To reinforce the point, here is another example: A single mutation in the homeotic gene known as *Apetala1* causes male floral reproductive structures (anthers) to form where petals are supposed to form in the flowers of field mustard, *Brassica oleracea* (Figure 17.19). At least in the laboratory, such abundantly anthered mutants are exceptionally fertile plants.

Another example: The embryos of some vertebrate lineages are alike in the early stages of development. Their tissues form in similar ways when cells divide, differentiate, and interact. The gut and heart, bones, skeletal muscles, and other parts start to grow and develop in orderly spatial patterns that are strikingly similar among these groups.

How, then, did adults of different groups get to be so different? We can expect that heritable changes in the onset, rate, or completion of developmental steps led to many of the differences. Some changes could have increased or decreased relative sizes of tissues and organs. Some changes could have ended growth during a juvenile stage; the adults of certain species still do have some juvenile features.

Figure 17.19 How a single mutation in a homeotic gene in many plants influences flower form and function.

(**a**) Normal flower of field mustard (*Brassica oleracea*). A mutation in the *Apetala1* gene causes a badly distorted flower (**b**) to form. (**c**) Wild-type flower of common wall cress (*Arabidopsis thaliana*). Mutation of the *Apetala1* gene in this plant causes flowers with no petals to form (**d**).

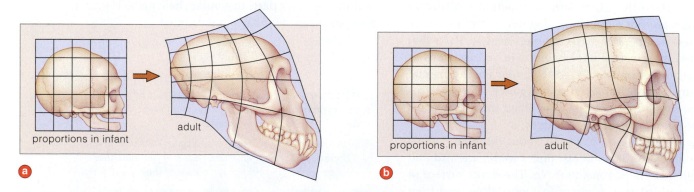

Figure 17.20 *Animated!* Differences between two primates, a possible outcome of mutations that changed the timing of steps in the body's development. The skulls are depicted as paintings on a rubber sheet divided into a grid. Stretching both sheets deforms the grid in a way that corresponds to differences in growth patterns between these primates. (**a**) Proportional changes in chimpanzee skull, and (**b**) human skull.

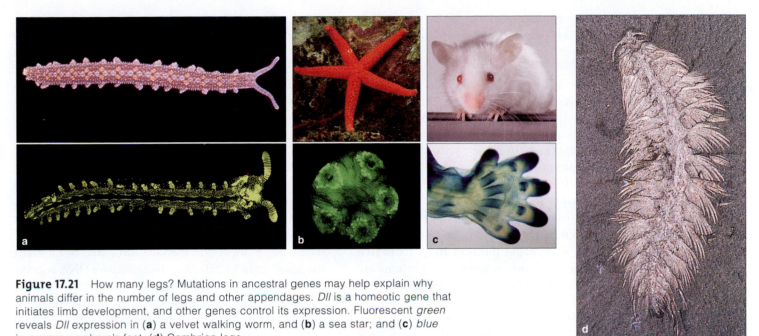

Figure 17.21 How many legs? Mutations in ancestral genes may help explain why animals differ in the number of legs and other appendages. *Dll* is a homeotic gene that initiates limb development, and other genes control its expression. Fluorescent *green* reveals *Dll* expression in (**a**) a velvet walking worm, and (**b**) a sea star; and (**c**) *blue* in a mouse embryo's foot. (**d**) Cambrian legs.

Modifications in genes that influence growth rates might have caused the major proportional differences between chimpanzee and human skull bones (Figure 17.20). For humans, the facial bones and skull bones around the brain increase in size at fairly consistent rates, from infant to adult. The rate of growth is faster for chimp facial bones, so the proportions of an infant skull and an adult skull differ significantly.

In addition, did transposons cause some variation among lineages? As Section 14.5 explains, these short DNA segments spontaneously and repeatedly slip into new places in genomes. Depending on where they end up, they can have powerful effects on gene expression.

Only primates carry the 300 base-pair transposons called *Alu* elements, and they have done so for at least 30 million years. *Alu* elements are noncoding, and yet they have sequences that resemble intron–exon splice signals. When inserted into coding regions of DNA, they promote duplications of themselves. *Alu* elements have had big effects on the expression of genes for estrogen, thyroid hormones, and other proteins that control growth and development. Were they pivotal in primate evolution? Possibly.

About 1 million *Alu* elements make up 10 percent of the human genome. The most recent shared ancestors of chimpanzees and humans diverged between 6 and 4 million years ago. More than 98 percent of human DNA is identical with chimpanzee DNA. Something in the remaining 2 percent accounts for the differences. Uniquely positioned *Alu* elements may be one factor.

As a final example, body appendages as different as crab legs, beetle legs, butterfly wings, sea star arms, fish fins, and mouse feet all start out as buds of tissue from the body surface. A bud will form wherever the *Dll* gene product is expressed. The product, a protein, is a signal for clusters of dividing embryonic cells to "stick out from the body" in an expected pattern, as in Figure 17.21. Normally, *Hox* genes help sculpt the body's details by suppressing expression of *Dll* where appendages are not supposed to form.

The *Dll* gene is expressed in similar ways across many phyla, which is strong evidence for its ancient origin. Indeed, in some Cambrian fossils, it looks like it was not suppressed at all (Figure 17.21*d*). Probably layers of gene controls evolved over time, resulting in the variable numbers and locations of appendages we observe today among all complex animals, including humans and other vertebrates.

Similarities and differences in patterns of development are often clues to shared ancestry, especially for the embryos of plants and animals.

Heritable changes that alter key steps in a developmental program may be enough to bring about major differences in the adult forms of related lineages.

Mutations in master genes and insertions of transposons into coding regions of DNA could have been enough to launch body plans in new evolutionary directions.

17.9 Clues in DNA, RNA, and Proteins

LINKS TO
SECTIONS
3.6, 14.4, 16.3, 16.4

All species are a mix of ancestral and novel traits, including biochemical ones. The kinds and numbers of traits they do or do not share are clues to relationships.

Each species has its own DNA base sequence, which encodes instructions for making RNAs and proteins (Sections 3.6 and 14.4). We can expect genes to mutate in any line of descent. The more recently two lineages have diverged, the less time each will have had to accumulate unique mutations. That is why RNA and proteins of closely related species are more similar than those of more distantly related ones.

Identifying biochemical similarities and differences among species is now rapid, thanks to the methods of automated gene sequencing (Section 16.3). Extensive sequence data for numerous genomes and proteins are compiled in internationally accessible databases. With such data, we know (for example) that 31 percent of the 6,000 genes of yeast cells have counterparts in our genome. So do 40 percent of the 19,023 roundworm genes and 50 percent of the fruit fly genes.

PROTEIN COMPARISONS

Similarities in amino acids can be used to decipher connections between species and to study why certain proteins are highly conserved. When two species have many proteins with similar or identical amino acid sequences, they probably are closely related. When sequences differ considerably, many mutations have been built in, which indicates that a long time passed since the two species shared a common ancestor.

A few essential genes have evolved very little; they are highly *conserved* across diverse species. One such gene encodes cytochrome *c*. Species that range from aerobic bacteria to humans must make this protein component of electron transfer chains. In humans, its primary structure consists of only 104 amino acids. Figure 17.22 shows the striking similarity between the entire amino acid sequences for cytochrome *c* from a yeast, a plant, and an animal. And think about this: The *entire* amino acid sequence of human cytochrome

c is identical with that of chimpanzee cytochrome *c*. It differs by merely 1 amino acid in rhesus monkeys, 18 in chickens, 19 in turtles, and 56 in yeasts. With this biochemical information in hand, would you predict that humans are more closely related to chimpanzees or to rhesus monkeys? Chickens or yeast?

NUCLEIC ACID COMPARISONS

Mutations that cause differences between species are dispersed through DNA's nucleotide sequences. Some regions of those sequences are unique to each lineage. Researchers use the regions to estimate evolutionary distances. They isolate and then compare DNA from the nuclei, mitochondria, and chloroplasts of different species (Figure 17.23). They also compare DNA regions that encode ribosomal RNA (rRNA).

Nucleic acid hybridization refers to base-pairing between DNA strands from different sources (Section 16.2). In the hybrid molecule, more hydrogen bonds form between matched bases than mismatched bases. Strands with more matches associate strongly with each other. The amount of heat required to separate two strands of a hybrid can be used as a comparative measure of their similarity. Why? It takes more heat to disrupt hybrid DNA of closely related species.

Evolutionary distances are still being measured by DNA–DNA hybridizations, although automated gene sequencing now gives faster, more quantifiable results. Remember DNA fingerprinting (Section 16.4)? DNA restriction fragments from different species can be compared after gel electrophoresis has separated them.

Mitochondrial DNA (mtDNA), which mutates fast, is used to compare individuals of eukaryotic species. In sexually reproducing species, it is inherited intact from one parent—usually the mother. Thus, changes between maternally related individuals probably were caused by mutations, not by genetic recombinations.

Computer programs quickly compare collections of DNA sequencing data. Comparative analyses either reinforce or invite modification of evolutionary trees based on morphological findings and the fossil record.

$^+$NH$_3$-gly asp val glu lys gly lys lys ile phe ile met lys cys ser gln cys his thr val glu lys gly gly lys his lys thr gly pro asn leu his gly leu phe gly arg lys thr gly gln ala pro gly tyr

$^+$NH$_3$-ala ser phe ser glu ala pro pro gly asn pro asp ala gly ala lys ile phe lys thr lys cys ala gln cys his thr val asp ala gly ala gly his lys gln gly pro asn leu his gly leu phe gly arg gln ser gly thr thr ala gly tyr

$^+$NH$_3$-thr glu phe lys ala gly ser ala lys lys gly ala thr leu phe lys thr arg cys leu gln cys his thr val glu lys gly gly pro his lys val gly pro asn leu his gly ile phe gly arg his ser gly gln ala glu gly tyr

Figure 17.22 *Animated!* Comparison of the primary structure of cytochrome *c* from a yeast (*top row*), wheat plant (*middle*), and primate (*bottom*). *Gold* highlights parts of the amino acid sequence that are identical in all three. The probability that such a pronounced molecular resemblance resulted by chance is extremely low. Cytochrome *c* is a vital component of electron transfer chains in cells. Its amino acid sequence has been highly conserved even in these three evolutionarily distant lineages.

RACCOON RED PANDA   GIANT PANDA

                                                    SPECTACLED   SLOTH   SUN   BLACK   POLAR   BROWN
                                                    BEAR   BEAR   BEAR   BEAR   BEAR   BEAR

DIVERGENCE
15–20 million years ago

DIVERGENCE
approximately
40 million years ago

Figure 17.23 Example of biochemical comparisons that help construct and refine evolutionary trees. This tree for red pandas, giant pandas, and brown bears was confirmed using mitochondrial DNA sequence comparisons. Recently, researchers analyzed sequence data from three mitochondrial DNAs and one intron. They also studied more potential relatives. They found that red pandas may be more closely related to skunks, weasels, and otters than to raccoons.

However, gene transfers between species can slant the results. For example, after hybridization between two different species of plants, hybrid offspring may cross back to either parental species, thus transferring genes from one species into the other. Gene swapping is rampant among prokaryotic species.

MOLECULAR CLOCKS

Some researchers estimate the timing of divergence by comparing the numbers of neutral mutations in genes that have been highly conserved in different lineages. Because the mutations have little or no effect on the individual's survival or reproduction, we can expect that neutral mutations have accumulated in conserved genes at a fairly constant rate.

The accumulation of neutral mutations to the DNA of a lineage has been likened to the predictable ticks of a **molecular clock**. Turn the hands of such a clock back, so that the total number of ticks unwinds down through past geologic intervals. Where does the last tick stop? Theoretically, it stops close to the time when molecular, ecological, and geographic events put the lineage on its unique evolutionary road.

How are molecular clocks calibrated? The number of differences in DNA base sequences or amino acid sequences between species can be plotted against a series of branch points that researchers have inferred from the fossil record. Section 19.6 will show how this is done. Such diagrams may reflect relative times of divergences among species, phyla, and other groups.

Biochemical similarity is greatest among the most closely related species and smallest among the most remote.

ala asn lys asn lys gly ile ile trp gly glu asp thr leu met glu tyr leu glu asn pro lys lys tyr ile pro gly thr lys met ile phe val gly ile lys lys lys glu glu arg ala asp leu ile ala tyr leu lys lys ala thr asn glu-COO⁻

ala asn lys asn lys ala val glu trp glu glu asn thr leu tyr asp tyr leu leu asn pro lys lys tyr ile pro gly thr lys met val phe pro gly leu lys lys pro gln asp arg ala asp leu ile ala tyr leu lys lys ala thr ser ser-COO⁻

ala asn ile lys lys asn val leu trp asp glu asn asn met ser glu tyr leu thr asn pro lys lys tyr ile pro gly thr lys met ala phe gly gly leu lys lys glu lys asp arg asn asp leu ile thr tyr leu lys lys ala cys glu-COO⁻

Summary

Section 17.1 Awareness of evolution, or heritable changes in lines of descent, emerged long ago through biogeography, geology, and comparative morphology.

Section 17.2 Prevailing cultural belief systems influence our interpretation of natural events. In the nineteenth century, naturalists worked to reconcile traditional belief systems with a growing body of physical evidence in support of evolution.

Biology⊗Now
Read the InfoTrac article "Typecasting a Bit Part," Stephen J. Gould, The Sciences, March 2000.

Section 17.3 Charles Darwin and Alfred Wallace proposed a novel theory that natural selection can bring about evolution. Here are the theory's main premises:

Any population tends to grow in size until resources dwindle. Its individuals must compete more for them.

Individuals with forms of traits that make them more competitive tend to produce more offspring.

Over the generations, more competitive forms of traits that have a heritable basis increase in frequency in the population relative to less competitive forms.

Thus nature "selects" variations in traits that are more effective at helping individuals survive and reproduce; such traits are more adaptive in a given environment.

Biology⊗Now
Read the InfoTrac article "What Darwin's Finches Can Teach Us About the Evolutionary Origin and Regulation of Biodiversity," B. Rosemary Grant and Peter Grant, Bioscience, March 2003.

Section 17.4 Fossils are stone-hard evidence of life in the distant past. Many fossils are embedded in stacked layers of sedimentary rock. Generally, the oldest layers are near the bottom of the sequence and more recently deposited layers are on top. Although the fossil record is not complete, it reveals much about life in the past.

Biology⊗Now
Learn more about fossil formation and the geologic time scale with the animations on BiologyNow.

Section 17.5 Fossil sequences were the basis for the first geologic time scale, which used abrupt transitions in the fossil record as boundaries for different eras. Through radiometric dating of fossils, absolute dates have since been assigned to the scale. This dating method has a relatively small margin of error.

Biology⊗Now
Learn more about the half-life of a radioisotope's atoms with the animated interaction on BiologyNow.

Section 17.6 The global distribution of land masses and fossils, magnetic patterns in volcanic rocks, and evidence of seafloor spreading from mid-oceanic ridges support the plate tectonic theory. According to this theory, slow movements of Earth's crustal plates raft land masses to new positions. Such movements had profound impacts on the directions of life's evolution.

Biology⊗Now
Learn more about drifting continents with the interaction on BiologyNow.

Section 17.7 Comparative morphology reveals evidence of evolution. Homologous structures are one of the clues. These body parts recur in different lineages, but they became modified in different ways after the lineages diverged from a shared ancestor. These parts are not the same as analogous structures. Such body parts did not start out being alike; they became similar in independent lineages as a response to similar kinds of environmental pressures.

Section 17.8 Similarities in patterns and structures of embryonic development suggest common ancestry. Even minor genetic changes can alter the onset, rate, and completion time of developmental stages. They can have major impact on the adult form.

Biology⊗Now
Explore proportional changes in embryonic development with the animated interaction on BiologyNow.

Section 17.9 We are now clarifying evolutionary relationships through comparisons of DNA, RNA, and proteins between different species. The investigative methods include nucleic acid hybridization and, more recently, automated gene sequencing and DNA fingerprinting, as explained in Chapter 16.

Some researchers estimate the times of divergences from ancestral lineages by comparing the number of neutral mutations in highly conserved genes. Such mutations alter the base sequence in DNA, but flexibility built into the genetic code keeps the change from altering the amino acid sequence of the specified protein. They may accumulate in the DNA of a species at a constant rate, like ticks of a molecular clock.

Biology⊗Now
Learn more about amino acid comparisons with the interaction on BiologyNow.

Self-Quiz   *Answers in Appendix II*

1. Biogeographers deal with _____ .
   a. patterns in which continents drift
   b. patterns in the world distribution of species
   c. mainland and island biodiversity
   d. both b and c are correct
   e. all are correct

2. _____ have influenced the fossil record.
   a. Sedimentation and compaction
   b. Crustal plate movements
   c. Volcanic ash deposition
   d. a through c

3. Life originated in the _____ eon.
   a. Archean   c. Phanerozoic
   b. Proterozoic   d. Cambrian

4. Which of these supercontinents formed first: Pangea or Gondwana?

5. Through _____ , the same body parts became modified differently in different lines of descent from a common ancestor.
   a. morphological convergence
   b. morphological divergence
   c. ancestral analogy
   d. ancestral homology

6. Homologous structures among major groups of organisms may differ in _____ .
   a. size   c. function
   b. shape   d. all of the above

7. By altering steps in the program by which embryos develop, _____ may lead to major differences between adults of related lineages.
   a. automated gene sequencing   c. transposons
   b. homeotic gene mutations   d. b and c

8. Molecular clocks are based on comparisons of _____ mutations in _____ genes.
   a. beneficial; moderately conserved
   b. neutral; moderately conserved
   c. neutral; highly conserved
   d. lethal; highly conserved

9. Match the terms with the most suitable description.
   ____ stratification   a. evidence of life in distant past
   ____ fossils   b. theory of repetitive change
   ____ homeotic   only in Earth history
        genes   c. body parts of similar size,
   ____ half-life   shape, or function in different
   ____ homologous   lineages with shared ancestor
        structures   d. insect wing and bird wing
   ____ uniformity   e. time it takes to decay half of a
   ____ analogous   quantity of a radioisotope's
        structures   atoms into something else
                         f. big role in development
                         g. layers of sedimentary rock

Additional questions are available on **Biology⟨S⟩Now**™

Critical Thinking

1. At one time, all species were ranked in a great Chain of Being, from lowly forms to Man, then to spiritual beings. Even some modern scientists still call the traits of species "rudimentary" or "advanced." Does the theory of natural selection imply that all species become more complex over time? Why or why not?

2. At the end of your backbone is a coccyx, a few small, fused-together bones (Figure 17.4). Is the human coccyx a *vestigial* structure—all that is left of the tail of some distant vertebrate ancestors? Or is it the start of a newly evolving structure?
   Formulate a hypothesis, then design a way to test the predictions you make based on the hypothesis.

3. Think about the species living around you. From the evolutionary perspective, which ones are most successful in terms of sheer numbers, geographic distribution, and how long their lineage has endured on Earth?

4. Comparative biochemistry can help us estimate evolutionary relationship and approximate times for divergences from ancestral stocks. Base sequence

Figure 17.24 Reconstruction of *Rodhocetus* based on fossils discovered in Pakistan. This cetacean lived 47 million years ago, along the shores of the Tethys Sea. Its ankle bones are strong evidence of a close evolutionary link between early whales and hoofed land mammals. Compare Figure 17.4.

comparisons and amino acid comparisons yield good estimates. Reflect on the genetic code (Section 14.2), then suggest why it may be a useful measure of mutations, mutation rates, and biochemical relatedness.

5. For some time, evolutionists accepted that the ancestors of whales were four-legged animals that walked on land, then took up life in water about 55 million years ago. Fossils reveal gradual changes in skeletal features that made an aquatic life possible. But which four-legged mammals were the ancestors?
   The answer recently came from Philip Gingerich and Iyad Zalmout. While digging for fossils in Pakistan, they found remains of early aquatic whales. Intact, sheeplike ankle bones *and* archaic whale skull bones were in the same fossilized skeletons (Figures 17.4 and 17.24).
   Ankle bones of fossilized, early whales from Pakistan have the same form as the unique ankle bones of extinct and modern artiodactyls. Modern cetaceans no longer have even a remnant of an ankle bone. Here is evidence of an evolutionary link between certain aquatic mammals and a major group of mammals on land.
   The radiometrically dated fossils are real. Yet no one was around to witness this transitional time. Because we did not see ancient life evolving, do you think there can be absolute proof of evolution in the distant past? Is the circumstantial evidence of fossil morphology enough to convince you that the theory is not wrong?

Rise of the Super Rats

Slipping in and out of the pages of human history are rats—*Rattus*—the most notorious of mammalian pests. One kind of rat or another has distributed pathogens and parasites that cause bubonic plague, typhus, and other deadly infectious diseases (Figure 18.1). The death toll from fleas that bit infected rats and then bit people has exceeded the dying in all wars combined.

The rats themselves are far more successful. By one estimate, there is one rat for every person in urban and suburban centers of the United States. Besides spreading diseases, rats chew their way through walls and wires of homes and cities. In any given year, they cause economic losses approaching 19 billion dollars.

For years, people have been fighting back with traps, ratproof storage facilities, and poisons, including arsenic and cyanide. During the 1950s, they used baits laced with warfarin. This synthetic organic compound interferes with blood clotting. Rats ate the baits. They died within days after bleeding internally or losing blood through cuts or scrapes.

Warfarin was extremely effective. Compared to other rat poisons, it had a lot less impact on harmless species. It quickly became the rodenticide of choice.

In 1958, however, a Scottish researcher reported that warfarin did not work against some rats. Similar reports from other European countries followed. About twenty years later, 10 percent of the urban rats caught in the United States were warfarin resistant. *What happened?* To find out, researchers compared warfarin-resistant rat populations with still-vulnerable rats. They traced the difference to a gene on one of the rat chromosomes.

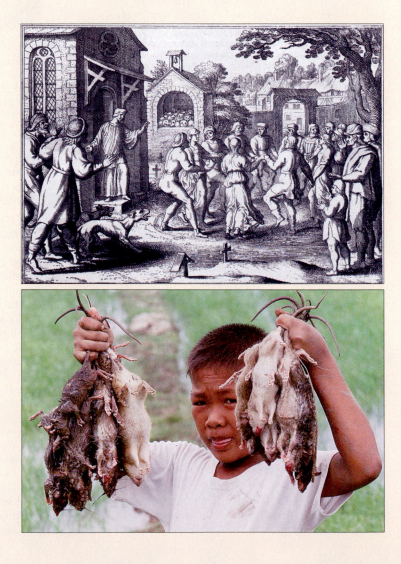

Figure 18.1 *Above*, medieval attempts to deal with a bubonic plague pandemic—the Black Death—that may have killed half the people in Europe alone. Not knowing that the disease agent hitches rides on rats, Europeans tried to protect themselves by praying and dancing until they dropped. Physicians wore bird masks, such as the mask shown on the facing page. They filled the "beak" with herbs that supposedly purified the air that plague victims had breathed. For the next 300 years, anyone accused of causing an outbreak of the plague, no matter how absurd the evidence, was burned alive.

Below, example of rats in this century. Rats infest 80,000 hectares of the rice fields in the Philippine Islands. They ruin more than 20 percent of the annual crops. Rice is the main food source for people in Southeast Asia.

Today we douse agricultural land and buildings with ever more potent rat poisons. By doing so, we have unwittingly promoted the rise of super rats. Three centuries from now, how will people be viewing *our* actions?

At that gene locus, a dominant allele was common among the warfarin-resistant rat populations but rare among the vulnerable ones. The dominant allele's product actually neutralizes warfarin's effect on blood clotting.

"What happened" was evolution by natural selection. As warfarin started to exert pressure on rat populations, the rat populations changed. The previously rare dominant allele suddenly proved to be adaptive. The lucky rats that inherited the allele survived and produced more offspring. The unlucky ones that inherited the recessive allele had no built-in defense, and they died. Over time, the dominant allele's frequency increased in all rat populations exposed to the poison.

Selection pressures can and often do change. When warfarin resistance increased in rat populations, people stopped using warfarin. Not surprisingly, the frequency of the dominant allele declined. Now the latest worry is the evolution of "super rats," which the newer and even more potent rodenticides cannot seem to kill.

The point is, when you hear someone question whether life evolves, remember this: With respect to life, **evolution** simply means heritable change is occurring in some line of descent. The actual mechanisms that can bring about such change are the focus of this chapter. Later chapters highlight how these mechanisms have contributed to the evolution of new species.

Watch the video online!

☑ How Would You Vote?

Antibiotic-resistant strains of bacteria are becoming dangerously pervasive. Standard animal husbandry practice includes continually dosing healthy animals with antibiotics—the same antibiotics prescribed for people. Should this practice stop? See BiologyNow for details, then vote online.

Key Concepts

WHAT IS MICROEVOLUTION?

Individuals of all natural populations share a gene pool but differ in which alleles they inherit. As a result, they show variations in phenotypes.

An individual does not evolve. Rather, a *population* evolves, which means its shared pool of alleles is changing. Over the generations, any allele may increase in its frequency among individuals, or it may become rare or lost.

Microevolution refers to changes in allele frequencies as an outcome of mutation, natural selection, genetic drift, and gene flow. Sections 18.1, 18.2

NATURAL SELECTION

Natural selection is the outcome of variation in heritable traits that influence which individuals of a population survive and reproduce in each generation. Selective agents operating in the environment can stabilize, disrupt, or cause directional shifts in the range of variation. Sections 18.3–18.6

GENETIC DRIFT

Sometimes chance events bring about random changes in allele frequencies over time. The magnitude of this genetic drift is greatest in small populations, where it can lead to a loss of genetic diversity. Section 18.7

GENE FLOW

Gene flow is the physical movement of alleles into and out of a population. It tends to oppose the effects of mutation, natural selection, and genetic drift; it keeps populations of a species similar to one another. Section 18.8

ADAPTATION AND THE ENVIRONMENT

An evolutionary adaptation is a heritable aspect of form, function, behavior, or development that contributes to the fit between an individual and its environment. The challenge is to identify environmental conditions to which a given trait is presumably adapted. Section 18.9

Links to Earlier Concepts

Before starting this chapter, review the premises of the theory of natural selection as outlined in Sections 1.4 and 17.3 as well as the definitions of basic terms in genetics (11.1).

You will be drawing upon your knowledge of mutation (14.5) and the chromosomal basis of inheritance (12.5 especially). We urge you to scan earlier sections on causes of continuous variation in populations (11.5) and on how the environment can modify gene expression (11.6).

18.1 Individuals Don't Evolve, Populations Do

LINKS TO
SECTIONS 11.4,
11.6, 11.7, 14.5, 17.9

*As Charles Darwin and Alfred Wallace perceived long ago, individuals don't evolve; populations do. Each **population** is a group of individuals of the same species in a specified area. To understand how it evolves, start with variation in the traits that characterize it.*

VARIATION IN POPULATIONS

The individuals of a population share certain features. Pigeons have two feathered wings, three toes forward, one toe back, and so on. These are *morphological* traits (*morpho–*, form). The individuals share *physiological* traits, including metabolic activities that help the body function in the environment. They respond the same way to certain basic stimuli, as when babies imitate adult facial expressions. These are *behavioral* traits.

However, the individuals of a population also show variation in the details of the shared traits. You know this just by thinking about the variations in the color and patterning of pigeon feathers or butterfly wings or snail shells. Figure 18.2 only hints at the range of variations in human skin color and distribution, color, texture, and amount of hair. Almost every trait of any species may vary, but variation can be dramatic among sexual reproducers.

For sexually reproducing species, at least, we may define the population as a group of individuals that are interbreeding, that are reproductively isolated from other species, and that produce fertile offspring. The offspring typically have two parents, and they have mixes of the parental forms of traits.

Many traits show *qualitative* differences; they have two or more distinct forms, or morphs. Remember the purple or white pea plant flowers that Gregor Mendel studied? The persistence of two forms of a trait in a population is a case of **dimorphism**. The persistence of three or more forms is **polymorphism**. In addition, for many traits, the individuals of a population show *quantitative* differences, a range of incrementally small variations in a specified trait (Section 11.7).

THE GENE POOL

Genes encode information about heritable traits. The individuals of a population inherit the same number and kind of genes (except for a pair of nonidentical sex chromosomes). Together, they and their offspring represent a **gene pool**—a pool of genetic resources.

For sexual reproducers, nearly all genes available in the shared pool have two or more slightly different molecular forms, or **alleles**. Any individual might or might not inherit identical alleles for any trait. This is

the source of variations in *phenotype*, or differences in the details of shared traits. Whether you have black, brown, red, or blond hair depends upon which alleles you inherited from your two parents.

You read about the inheritance of alleles in earlier chapters. Here we summarize the key events involved:

Gene mutation

Crossing over at meiosis I (puts novel combinations of alleles in chromosomes)

Independent assortment at meiosis I (puts mixes of maternal and paternal chromosomes in gametes)

Fertilization (combines alleles from two parents)

Change in chromosome number or structure (loss, duplication, or repositioning of genes)

Only mutation creates new alleles. The other events shuffle existing alleles into different combinations, but what a shuffle! Each gamete gets one of many millions of possible combinations of maternal and paternal chromosomes that may or may not be identical at each locus. Unless you are an identical twin, it is extremely unlikely that another person with your precise genetic makeup has ever lived or ever will.

One other point about the nature of the gene pool: Offspring do not inherit phenotypes; they inherit *genes*. Section 11.6 describes how environmental conditions, too, bring about variation in the range of phenotypes, but the effects last no longer than the individual.

MUTATION REVISITED

Being the original source of new alleles, mutations are worth another look—this time in the context of their impact on populations. Usually, gene mutations that have beneficial or neutral effects are transmitted to a new generation. We cannot predict precisely when or in which individual a particular gene will mutate. We *can* predict rates of mutation, or the probability that a mutation will happen in a specified interval (Section 14.5). For instance, one estimated rate for mammalian genomes is 2.2^{-9} mutations per base pair per year.

Many mutations give rise to structural, functional, or behavioral alterations that reduce an individual's chances of surviving and reproducing. Even a single biochemical change may be devastating. For instance, skin, bones, tendons, lungs, blood vessels, and many other vertebrate organs incorporate collagen. Thus, when the collagen gene has mutated, drastic problems may ripple all through the body. Compare Section 11.4.

Any mutation that results in severe disruptions in phenotype usually causes death. It is a **lethal mutation**.

A **neutral mutation**, recall, alters the base sequence in DNA, but the change has no discernible effect on survival or reproduction (Section 17.9). It neither helps nor hurts the individual. For instance, if you carry a mutant gene that keeps your earlobes attached to the head instead of swinging freely, this in itself should not stop you from surviving and reproducing as well as anybody else. Therefore, natural selection does not affect the frequency of the trait in the population.

Every so often, a mutation proves useful. A mutant gene product that affects growth might make a corn plant grow larger or faster and thereby give it the best access to sunlight and nutrients. A neutral mutation might prove helpful if conditions in the environment change. Even if a mutant gene bestows only a slight advantage, natural selection or a chance event might favor its preservation in DNA and its transmission to the next generation.

Mutations are rare, so they usually have little or no immediate effect on a population's allele frequencies. But they have been slipping into genomes for billions of years. Cumulatively, they have served as reservoirs for change, for biodiversity that is staggering in its breadth. Think of it. The reason you don't look like a bacterium or an avocado or earthworm or even your neighbors down the street began with mutations that arose at different times, in different lines of descent.

STABILITY AND CHANGE IN ALLELE FREQUENCIES

Researchers typically track **allele frequencies**, or the relative abundances of alleles of a given gene among all individuals of a population. They can start from a theoretical reference point, **genetic equilibrium**, when a population is *not* evolving with respect to that locus.

Genetic equilibrium can only occur if five conditions are being met: There is no mutation, the population is infinitely large, the population is isolated from other populations of the same species, individuals mate at random, and all individuals survive and produce the same number of offspring.

If you are interested, the following section offers a closer look at the nature of genetic equilibrium—the point at which a population is not evolving.

As it happens, genetic equilibrium is exceedingly rare in nature. Why? Mutations are rare but inevitable, and they might throw a wild card in the game of who survives and reproduces. Also, three processes—called *natural selection, genetic drift,* and *gene flow*—can drive populations out of equilibrium. **Microevolution** refers to small-scale changes in allele frequencies that arise as an outcome of mutation, natural selection, genetic drift or gene flow, or some combination of these.

Figure 18.2
A sampling of the phenotypic variation in populations of humans and snails, the outcome of variations in frequencies of alleles.

We partly characterize a natural population or species by morphological, physiological, and behavioral traits, most of which are heritable.

At any gene locus, different alleles give rise to variations in individual phenotypes—to differences in the details of shared structural, functional, and behavioral traits.

The individuals of a population share a pool of genetic resources—that is, a pool of alleles.

Only mutation creates new alleles. Natural selection, genetic drift, and gene flow affect only the frequencies of various alleles at a given gene locus in the population.

Most populations are slowly evolving, which simply means that the frequencies of the alleles for a specified trait are changing from one generation to the next.

18.2   When Is A Population *Not* Evolving?

How do researchers know whether or not a population is evolving? They can start by tracking deviations from the baseline of genetic equilibrium.

The Hardy–Weinberg Formula   Early in the twentieth century, Godfrey Hardy (a mathematician) and Wilhelm Weinberg (a physician) independently applied the rules of probability to sexually reproducing populations. Like the geneticists who came after them, they perceived that gene pools can remain stable only when five conditions are being met:

1. There is no mutation.

2. The population is infinitely large.

3. The population is isolated from all other populations of the species (no gene flow).

4. Mating is random.

5. All individuals survive and produce the same number of offspring.

In other words, allele frequencies for any gene in the shared pool will remain stable unless the population is evolving. Hardy and Weinberg developed a simple formula that can be used to track whether a population of any sexually reproducing species is slipping out of that state of genetic equilibrium.

Consider tracking a hypothetical pair of alleles that affect butterfly wing color. A protein pigment is specified by dominant allele A. If a butterfly inherits two AA alleles, it will have dark-blue wings. If it inherits two recessive alleles aa, it will have white wings. If it inherits one of each (Aa), the wings will be medium-blue (Figure 18.3).

At genetic equilibrium, the proportions of the wing-color genotypes are

$$p^2(AA) + 2pq(Aa) + q^2(aa) = 1.0$$

where *p* and *q* are the frequencies of alleles *A* and *a*. This is what became known as the *Hardy–Weinberg equilibrium equation*. It defines the frequency of a dominant and a recessive allele for a gene that controls a particular trait in a population.

The frequencies of *A* and *a* must add up to 1.0. To give a specific example, if *A* occupies half of all the loci for this gene in the population, then *a* must occupy the other half (0.5 + 0.5 = 1.0). If *A* occupies 90 percent of all the loci, then *a* must occupy 10 percent (0.9 + 0.1 = 1.0). No matter what the proportions,

$$p + q = 1.0$$

At meiosis, recall, paired alleles segregate and end up in different gametes. So the proportion of gametes having the *A* allele is *p*. The proportion having the *a* allele is *q*. The Punnett square on the next page reveals the genotypes possible in the next generation (AA, Aa, and aa).

490 *AA* butterflies dark-blue wings   490 *AA* butterflies dark-blue wings   490 *AA* butterflies dark-blue wings

420 *Aa* butterflies medium-blue wings   420 *Aa* butterflies medium-blue wings   420 *Aa* butterflies medium-blue wings

90 *aa* butterflies white wings   90 *aa* butterflies white wings   90 *aa* butterflies white wings

Starting Population   **Next Generation**   **Next Generation**

Figure 18.3 *Animated!* How to determine whether a population is evolving. The frequencies of wing-color alleles among all individuals in this hypothetical population of morpho butterflies have not changed because all five assumptions upon which the Hardy–Weinberg rule is based are being met.

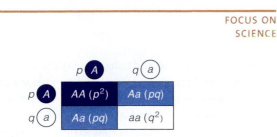

The frequencies add up to 1.0: $p^2 + 2pq + q^2 = 1.0$.

Suppose that the population has 1,000 individuals and that each one produces two gametes:

490 *AA* individuals make 980 *A* gametes
420 *Aa* individuals make 420 *A* and 420 *a* gametes
90 *aa* individuals make 180 *a* gametes

The frequency of alleles *A* and *a* among 2,000 gametes is

$$A = \frac{980 + 420}{2,000 \text{ alleles}} = \frac{1,400}{2,000} = 0.7 = p$$

$$a = \frac{180 + 420}{2,000 \text{ alleles}} = \frac{600}{2,000} = 0.3 = q$$

At fertilization, gametes combine at random and start a new generation. If the population size is still 1,000, you will find 490 *AA*, 420 *Aa*, and 90 *aa* individuals. Because the allele frequencies for dark-blue, medium-blue, and white wings are the same as they were in the original gametes, they will give rise to the same phenotypic frequencies that occurred in the preceding generation.

As long as the assumptions that Hardy and Weinberg identified continue to hold, the pattern will persist. If traits show up in different proportions from one generation to the next, however, then one or more of the five assumptions is not being met. The hunt can begin for one or more of the evolutionary forces driving the change.

Applying the Rule   So how does the Hardy–Weinberg formula work in the real world? For one thing, researchers use it to estimate the frequency of carriers of alleles that cause genetic traits and disorders.

For example, about 1 percent of people of Irish ancestry are affected by *hemochromatosis*. They absorb too much iron from their food. Symptoms of this autosomal recessive disorder include liver problems, fatigue, and arthritis. We can use the number to estimate the frequency of carriers of the recessive allele. If $p^2 = 0.01$, then p is 0.1, q is 0.9, and the carrier frequency ($2pq$) must be 0.18 among Irish populations. Such information is useful to doctors and public health professionals.

Another example: A deviation from the frequencies predicted by the Hardy–Weinberg formula suggests that a mutant allele for *BRCA2* may be lethal to female embryos. The allele also has been linked to breast cancer. For one study, researchers tracked the frequency of the mutant allele among newborn girls. There were fewer homozygotes than expected, based on the number of heterozygotes and the Hardy–Weinberg formula. By itself or in combination with other alleles, a pair of mutant *BRCA2* alleles may cause the spontaneous abortion of the early embryo.

18.3   Natural Selection Revisited

Natural selection, again, is the outcome of differences in reproduction among individuals of a population that vary in their shared traits, some of which prove more adaptive than others under prevailing environmental conditions.

LINKS TO
SECTIONS
1.4, 17.3

Natural selection may be the most influential process of microevolution. Its impact shows up at all levels of biological organization, which is the reason you were introduced to it early on, in Chapter 1. You also came across simple examples in other chapters, and Sections 17.2 and 17.3 offered you a glimpse of the history that preceded its discovery. Turn now to major categories of selection, as sketched out in Figure 18.4.

With *directional* selection, the range of variation for a trait shifts in a consistent direction; individuals at one end of the range of variation are selected against and those at the other end are favored. With *stabilizing* selection, the forms at one or both ends of the range are selected against. With *disruptive* selection, forms at one or both ends are favored and intermediate forms are selected against.

Diverse selection pressures acting on a population might favor forms at one end in the range of variation for a trait, or intermediate forms within that range, or extreme forms at both ends of the range.

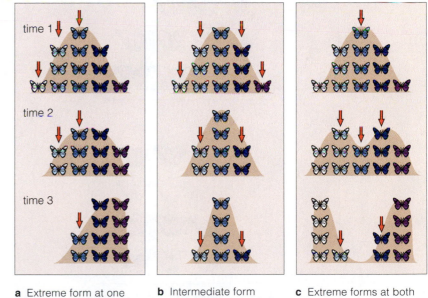

a Extreme form at one end of the range of phenotypes favored

b Intermediate form of the range of phenotypes favored

c Extreme forms at both ends of the range of phenotypes favored

Figure 18.4   Overview of the outcomes of three modes of natural selection: (**a**) directional, (**b**) stabilizing, and (**c**) disruptive.

18.4 Directional Selection

LINKS TO
SECTIONS
1.4, 16.7

*With **directional selection**, allele frequencies shift in a consistent direction, so forms at one end of a phenotypic range become more common than midrange forms, as in Figure 18.5. Directional change in the environment or novel conditions can cause the shift.*

RESPONSES TO PREDATION

The Peppered Moth Populations of peppered moths (*Biston betularia*) offer us a classic case of directional selection. The moths feed and mate at night and rest motionless on trees during the day. Their behavior and coloration (mottled gray to nearly black) camouflage them from day-flying, moth-eating birds.

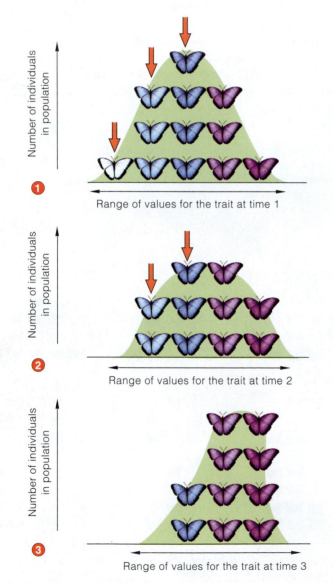

Figure 18.5 *Animated!* Directional selection. These bell-shaped curves signify a range of continuous variation in a butterfly wing-color trait. *Medium-blue* is between two phenotypic extremes—*white* and *dark purple*. Orange arrows signify which forms are being selected against over time.

In the 1850s, the industrial revolution started in England, and factory smoke altered conditions in much of the countryside. Before then, light moths were the most common form, and a dark form was rare. Also, light-gray speckled lichens had grown thickly on tree trunks. Light moths but not dark moths that rested on the lichens were camouflaged (Figure 18.6a).

Lichens are sensitive to air pollution. Between 1848 and 1898, soot and other pollutants started to kill the lichens and darken tree trunks. The dark moth form was better camouflaged (Figure 18.6b). Researchers hypothesized: If the original conditions favored light moths, then the *changed* conditions favored dark ones.

In the 1950s, H. B. Kettlewell used a *mark–release–recapture method* to test the possibility. He bred both moth forms in captivity and marked hundreds so that they could be easily identified after being released in the wild. He released them near highly industrialized areas around Birmingham and near an unpolluted part of Dorset. His team recaptured more dark moths in the polluted area and more light ones near Dorset:

	Near Birmingham (pollution high)	Near Dorset (pollution low)
Light-Gray Moths		
Released	64	393
Recaptured	16 (25%)	54 (13.7%)
Dark-Gray Moths		
Released	154	406
Recaptured	82 (53%)	19 (4.7%)

Observers also hid in blinds near moths that had been tethered to trees. They observed birds capturing more light moths around Birmingham and more dark ones around Dorset. Directional selection was in play.

Figure 18.6 Natural selection of two forms of the same trait, body surface coloration, in two settings. (**a**) Light moths (*Biston betularia*) on a nonsooty tree trunk are hidden from predators. Dark ones stand out. (**b**) The dark color is more adaptive in places where soot darkens tree trunks.

Figure 18.7 Visible evidence of directional selection in a population of rock pocket mice relative to a neighboring population, as documented by Michael Nachman, Hoi Hoekstra, and Susan D'Agostino. (**a**) Lava basalt flow at the study site. The two color morphs of rock pocket mice, each posed on two different backgrounds: (**b**) tawny fur and (**c**) dark fur.

Pollution controls went into effect in 1952. Lichens made a comeback, and tree trunks became largely free from soot. Phenotypes shifted in the reverse direction. Where pollution has decreased, the frequency of dark moths has been decreasing as well.

Pocket Mice Directional selection is at work among rock pocket mice (*Chaetodipus intermedius*) of Arizona's Sonoran Desert. Of more than eighty genes known to affect coat color in mice, researchers found a gene that governs a difference between two populations of this mouse species (Figure 18.7).

Rock pocket mice are small mammals that spend the day in underground burrows and forage for seeds at night. Some live in tawny-colored outcroppings of granite. In this habitat, individuals with tawny fur are camouflaged from predators (Figure 18.7b).

A smaller population of pocket mice lives in the same region, but these mice scamper over dark basalt of ancient lava flows. They have dark coats, so they, too, are camouflaged from predators (Figure 18.7c).

We can expect that night-flying predatory birds are selective agents that affect fur color. For instance, owls have an easier time seeing mice with fur that does not match the rocks.

Michael Nachman used genetic data on laboratory mice to formulate a hypothesis on differences in coat color in the two wild populations of pocket mice. He predicted that a mutation of either the *Mclr* gene or *agouti* gene could cause the difference. He collected DNA from dark pocket mice at a lava flow and from light mice at adjacent granite outcroppings.

DNA analysis showed that the *Mclr* gene sequence for all dark mice differed by four nucleotides from that of their light-furred neighbors. In the population of dark mice, the allele frequencies had evolved in a consistent direction as a result of selection pressure, so dark fur became more common.

RESISTANCE TO PESTICIDES AND ANTIBIOTICS

Pesticides can cause directional selection, as they did for the super rats. Typically, a heritable aspect of body form, physiology, or behavior helps a few individuals survive the first pesticide doses. As the most resistant ones are favored, resistance becomes more common. About 450 species of pests are now resistant to one or more types of pesticides. Also, some pesticides kill off the natural predators. Freed from natural constraints, resistant populations flourish and inflict more damage. This result of directional selection is *pest resurgence*. Some genetically engineered crop plants resist pests. In time, they too may exert selection pressure.

Antibiotics also can result in directional selection. Certain microbes produce natural antibiotics that can kill bacterial competitors for nutrients. We use natural and synthetic antibiotics to fight pathogenic bacteria. Streptomycins, for example, inhibit protein synthesis in bacterial cells. The penicillins disrupt covalent bonds that hold a bacterial cell wall together.

Yet antibiotics have been overprescribed, often for simple infections that would clear up on their own. Genetic variation in bacterial gene pools allows some cells with certain genotypes to survive as others die. So overuse of antibiotics favors the resistant bacterial populations, which will be harder to eradicate in the millions of people who contract cholera, tuberculosis, and other bacterial diseases each year. Also, healthy farm animals are routinely dosed with antibiotics to prevent infection. Consider: In eggs that look slightly fluorescent green, tetracycline is showing through.

With directional selection, allele frequencies underlying a range of variation tend to shift in a consistent direction in response to some change in the environment.

18.5 Selection Against Or in Favor of Extreme Phenotypes

LINK TO
SECTION
17.3

Consider now two more categories of natural selection. One works against phenotypes at the fringes of a range of variation; the other favors them.

STABILIZING SELECTION

With **stabilizing selection**, intermediate forms of a trait in a population are favored, and extreme forms are not. This mode of selection can counter mutation, genetic drift, and gene flow. It tends to preserve intermediate phenotypes in the population (Figure 18.8a).

As an example, prospects are not good for human babies who weigh far more or far less than average at birth. Also, pre-term instead of full-term pregnancies increase the risk, as Figure 18.9 indicates.

Newborns weighing less than 5.51 pounds or born before thirty-eight weeks of pregnancy are completed tend to develop high blood pressure, diabetes, and heart disease when they are adults. Researchers now suspect that the mother's blood concentration of a stress hormone, cortisol, is linked to low birth weight and illnesses that develop later in life.

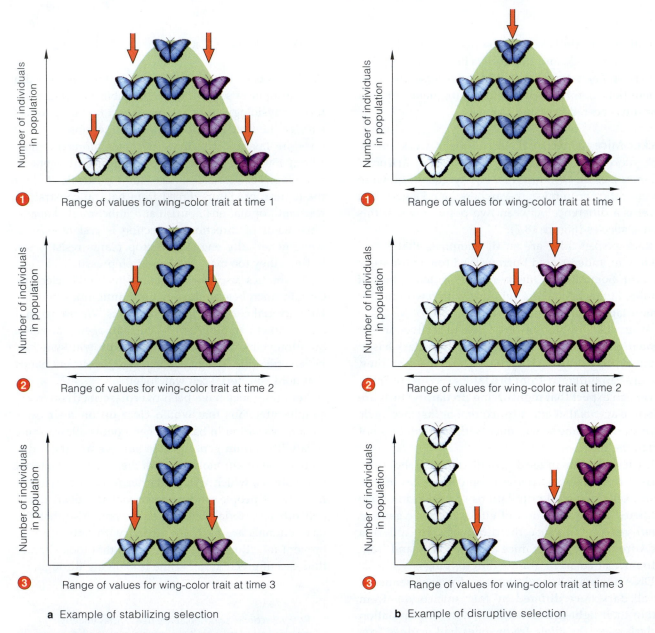

a Example of stabilizing selection

b Example of disruptive selection

Figure 18.8 *Animated!* Selection against or in favor of extreme phenotypes, with a population of butterflies as the example. (**a**) stabilizing selection and (**b**) disruptive selection. The *orange* arrows show forms of the trait being selected against.

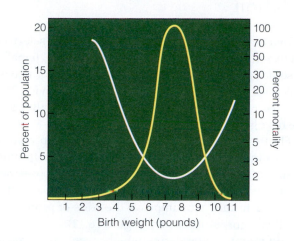

Figure 18.9 Weight distribution for 13,730 human newborns (*yellow* curve) correlated with death rate (*white* curve).

Figure 18.10 Adult sociable weaver (*Philetairus socius*), a native of the African savanna. These birds cooperate in constructing and using large communal nests in a region where trees and other good nesting sites are scarce.

Rita Covas and her colleagues gathered evidence of stabilizing selection on the body mass of juvenile and adult sociable weavers (*Philetairus socius*), as in Figure 18.10. Between 1993 and 2000, they captured, measured, tagged, released, and recaptured 70 to 100 percent of the birds living in communal nests during the breeding season. Their field studies supported a prediction that body mass is a trade-off between risks of starvation and predation. Intermediate-mass birds have the selective advantage. Foraging is not easy in this habitat, and lean birds do not store enough fat to avoid starvation. We can expect that fat ones are more attractive to predators and not as good at escaping.

lower bill 12 mm wide   lower bill 15 mm wide

Figure 18.11 Disruptive selection in African finch populations. Selection pressures favor birds with bills that are about 12 *or* 15 millimeters wide. The difference is correlated with competition for scarce food resources during the dry season.

DISRUPTIVE SELECTION

With **disruptive selection**, forms at both ends of the range of variation are favored and intermediate forms are selected against (Figure 18.8*b*).

Consider the black-bellied seedcracker (*Pyrenestes ostrinus*) of Cameroon. Females and males of these African finches have large or small bills—but no sizes in between (Figure 18.11). It is like everyone in Texas being four feet *or* six feet tall, with no one in between.

The pattern holds all through the geographic range. If unrelated to gender or geography, what causes it? If only two bill sizes persist, then disruptive selection may be eliminating birds with intermediate-size bills. Factors that affect feeding performance are the key. Cameroon's swamp forests flood in the wet season; lightning-sparked fires burn in the hot, dry season. Most plants are fire-resistant, grasslike sedges. One species produces hard seeds and the other, soft seeds.

Remember the bills of Galápagos finches (Section 17.3)? Here, also, the ability to crack hard seeds affects survival. All Cameroon seedcrackers prefer soft seeds,

but birds with large bills are better at cracking hard ones. In the dry season, the birds compete fiercely for scarce seeds. A scarcity of both types of seeds during recurring episodes of drought has a disruptive effect on bill size in the seedcracker population. Birds with intermediate sizes are being selected against, and now all bills are either 12 *or* 15 millimeters wide.

In these seedcrackers, bills of a particular size have a genetic basis. In experimental crosses between two birds with the two optimal bill sizes, all offspring had a bill of one size or the other, nothing in between.

With stabilizing selection, intermediate phenotypes are favored and extreme phenotypes at both ends of the range of variation are eliminated.

With disruptive selection, intermediate forms of traits are selected against and extreme forms in the range of variation are favored.

18.6 Maintaining Variation in a Population

Natural selection theory helps explain diverse aspects of nature, including male–female differences and the relationship between sickle-cell anemia and malaria.

SEXUAL SELECTION

The individuals of many sexually reproducing species show a distinct male or female phenotype, or **sexual dimorphism** (*dimorphos,* having two forms). Often the males are larger and flashier than females. Courtship rituals and male aggression are common.

These adaptations and behaviors seem puzzling. All take energy and time away from an individual's survival activities. Why do they persist if they do not contribute directly to survival? The answer is **sexual selection**. By this mode of natural selection, winners are the ones that are better at attracting mates and successfully reproducing compared to others of the population. The most adaptive traits help individuals defeat same-sex rivals for mates or are the ones most attractive to the opposite sex.

Figure 18.12 One male bird of paradise in a flashy courtship perhaps, the sexual colorful female. The for females, which function as selective agents. (Why do you suppose the females are drab-colored?)

outcome of sexual selection. This (*Paradisaea raggiana*) is engaged display. He caught the eye (and, interest) of the smaller, less males of this species compete fiercely

By choosing mates, a male or female is a selective agent acting on its own species. For example, females of some species shop among a congregation of males, which vary in appearance and courtship behavior. The selected males and the females pass on their alleles to the next generation.

Flashy body parts and behaviors show up among species in which males provide little or no help with raising offspring. The female apparently chooses her partner on the basis of observable signs of health and vigor. Such traits may improve the odds of producing healthy, vigorous offspring (Figure 18.12).

You might be wondering whether we can correlate genes with specific forms of sexual behavior. The sexual deception practiced by an Australian orchid is a case in point. The flowers of *Chiloglottis trapeziformis* attract male wasps by secreting a substance that is identical with a sex pheromone—which female wasps release to attract male wasps. Flowers get pollinated as males attempt to copulate with them.

This orchid is stingy. It gives a male wasp nothing in return, not a single drop of nectar, even though it is the orchid's exclusive pollinator. The female wasps are wingless. They hatch in soil. When males do not lift and carry them to a food source, they starve to death.

When *C. trapeziformis* puts out blooms, male wasps waste precious time and metabolic energy trying to find females. Evolutionary biologist Florian Schiestl has proposed that selection pressure is afoot for wasps that can produce a new sex pheromone, one that the orchid cannot duplicate.

This interaction exploits male wasps, but Wittko Francke thinks it might put pressure on their brains to evolve. In an orchid patch, the average tiny-brained male wasp copulates blindly with whatever smells right. It will try to copulate even with the head of a pin that has a few micrograms of pheromone sprayed on it. However, a few wasps with a slightly less robotic brain might be able to identify the females by other cues, such as visual ones. Alternatively, both species could face extinction, another pattern in nature.

SICKLE-CELL ANEMIA—LESSER OF TWO EVILS?

With *balancing* selection, two or more alleles of a gene are being maintained at relatively high frequencies in the population. Their persistence is called **balanced polymorphism** (*polymorphos,* having many forms). The allele frequencies might shift slightly, but often they return to the same values over the long term. We may see this balance when conditions favor heterozygotes. In some way, their nonidentical alleles for a given trait grant them higher fitness compared to homozygotes, which, recall, have identical alleles for the trait.

Consider the environmental pressures that favor an Hb^A/Hb^S pairing in humans. The Hb^S allele codes for a mutant form of hemoglobin, an oxygen-transporting protein in blood. Homozygotes (Hb^S/Hb^S) develop the genetic disorder *sickle-cell anemia* (Section 3.6).

The Hb^S frequency is highest in both tropical and subtropical regions of Asia and Africa. Often, Hb^S/Hb^S homozygotes die in their early teens or early twenties. Yet, in these same regions, heterozygotes (Hb^A/Hb^S) make up nearly a third of the population! Why is this combination maintained at such high frequency?

The balancing act is most pronounced in areas that, historically, have had the highest incidence of *malaria* (Figure 18.13). Mosquitoes transmit the parasitic agent of malaria, *Plasmodium*, to human hosts. The parasite multiplies in the liver and then in red blood cells. The target cells rupture and release new parasites during severe, recurring bouts of infection (Section 22.7).

It turns out that Hb^A/Hb^S heterozygotes are more likely to survive malaria than people who make only normal hemoglobin. Several survival mechanisms are possible. In heterozygotes, the infected cells take on a sickle shape under normal conditions. The abnormal shape marks them as targets for the immune system, which destroys them, along with the parasites inside. In addition, heterozygotes have one functioning Hb^A allele. Although they are not completely healthy, they still produce enough normal hemoglobin to prevent sickle-cell anemia. That is why heterozygotes are more likely to survive long enough to reach reproductive age, compared to Hb^S/Hb^S homozygotes.

In short, the persistence of the "harmful" Hb^S allele may be a matter of relative evils. Malaria has been a selective force for thousands of years in tropical and subtropical areas of Asia, the Middle East, and Africa. Through that time span, natural selection has favored the Hb^A/Hb^S combination in all of the malaria-ridden regions, because heterozygotes show more resistance to the disease. In such environments, the combination has proved to have more survival value than either the Hb^S/Hb^S or the Hb^A/Hb^A combination.

With sexual selection, some version of a gender-related trait gives the individual an advantage in reproductive success. Sexual dimorphism is one outcome of sexual selection.

In a population showing balanced polymorphism, natural selection is maintaining two or more alleles at frequencies greater than 1 percent over the generations.

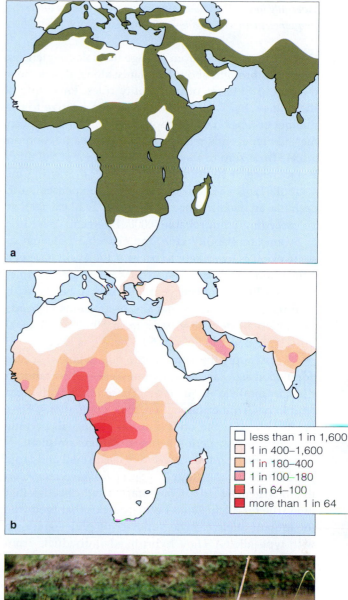

less than 1 in 1,600
1 in 400–1,600
1 in 180–400
1 in 100–180
1 in 64–100
more than 1 in 64

Figure 18.13 (**a**) Distribution of malaria cases reported in Africa, Asia, and the Middle East in the 1920s, before the start of programs to control mosquitoes, the vector for *Plasmodium*. (**b**) Distribution and frequency of people with the sickle-cell trait. Notice the close correlation between the maps. (**c**) Physician searching for *Plasmodium* larvae in Southeast Asia.

18.7 Genetic Drift—The Chance Changes

LINKS TO
SECTIONS
11.2, 12.10

Especially in small populations, random changes in allele frequencies can lead to a loss of genetic diversity.

Genetic drift is a random change in allele frequencies over time, brought about by chance alone. Researchers measure it in terms of probability rules. *Probability* is the chance that something will happen relative to the number of times it could happen (Section 11.2). We can measure an event's relative frequency as a fraction on a scale from zero to 1—or 0 to 100 percent of the time. For instance, if 10 million people enter a drawing for a month-long vacation in Hawaii, all expenses paid, each has an equal chance of winning: 1/10,000,000, or an exceedingly improbable 0.00001 percent.

By one probability rule, the expected outcome of some event is less likely to occur if the event happens only rarely. Each time you flip a coin, for example, there is a 50 percent chance it will turn up heads. With 10 flips, odds are high that the proportions of heads and tails will deviate greatly from 50:50. With 1,000 flips, large deviations from 50:50 are less likely.

We can apply the same rule to populations. Because population sizes are not infinite, there will be random changes in allele frequencies. These random changes tend to have minor impact on large populations. They greatly increase the odds that an allele will become more or less prevalent when populations are small.

Steven Rich and his coworkers used small and large populations of the flour beetle (*Tribolium castaneum*) to study genetic drift. They started with beetles that bred true for allele b^+ and other beetles that bred true for mutant allele b. (The superscript plus signifies a wild-type allele.) They hybridized individuals from both groups to get a population of F_1 heterozygotes

(b^+b), which they divided into sets of twelve. Different sets consisted of 10, 20, 50, and 100 randomly selected male and female beetles, and the subpopulation sizes were maintained for twenty generations.

Figure 18.14 shows two of the test results. Drift was greatest in the sets of 10 beetles and least in the sets of 100 beetles. Notice the loss of b^+ from one of the small populations (one graph line ends at 0 in Figure 18.14*a*). Only allele b remained. When all of the individuals of a population have become homozygous for one allele only at a locus, we say that **fixation** has occurred.

Thus, *random change in allele frequencies leads to the homozygous condition and a loss of genetic diversity over time.* This is genetic drift's outcome in all populations; it simply happens faster in small ones (Figure 18.14). Once alleles from the parent population have become fixed, their frequencies will not change again unless mutation or gene flow introduces new alleles.

BOTTLENECKS AND THE FOUNDER EFFECT

Genetic drift is pronounced when a few individuals rebuild a population or start a new one. This happens after a **bottleneck**, a drastic reduction in population size brought about by severe pressure. Suppose that contagious disease, habitat loss, or hunting nearly wipes out a population. Even if a moderate number of individuals survive a bottleneck, allele frequencies will have been altered at random.

In the 1890s, hunters killed all but twenty of a large population of northern elephant seals. Government restrictions allowed the population to recover to about 130,000 individuals. Each is homozygous for all of the genes analyzed so far.

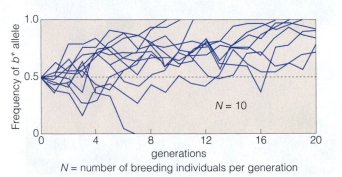

N = number of breeding individuals per generation

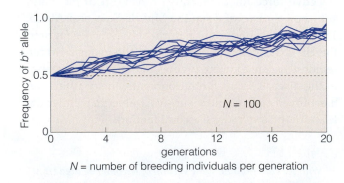

N = number of breeding individuals per generation

a The size of twelve populations of beetles was maintained at 10 breeding individuals per generation for twenty generations. Allele b^+ was lost and b became fixed in one population. Notice that alleles can be fixed or lost even in the absence of selection.

b The size of twelve populations was maintained at 100 individuals per generation for twenty generations. Allele b did not become fixed. Drift was far less in each generation than it was in the small populations tracked in (**a**).

Figure 18.14 *Animated!* Genetic drift's effect on allele frequencies in small and large populations. The starting frequency of mutant allele b^+ was 0.5.

Figure 18.15 Founder effect. This wandering albatross carries seeds, stuck to its feathers, from the mainland to a remote island. By chance, most of the seeds carry an allele for orange flowers that are rare in the original population. Without further gene flow or selection for color, genetic drift will fix the allele on the island.

Unpredictable genetic shifts can occur after a few individuals establish a new population. This form of bottlenecking is a **founder effect**. Genetic diversity might be greatly reduced relative to the original gene pool, as when a lone seed founds a population on a remote island in the middle of the ocean (Figure 18.15).

INBRED POPULATIONS

Genetic drift is less pronounced in inbred populations. **Inbreeding** is nonrandom mating among very close relatives, which share many identical alleles. It leads to the homozygous condition. It also lowers fitness if harmful recessive alleles are increasing in frequency.

Most human societies forbid or discourage incest (inbreeding between parents and children or siblings). Inbreeding among other close relatives is common in geographically or culturally isolated small groups. The Old Order Amish in Pennsylvania are moderately inbred. One outcome is a rather high frequency of a recessive allele that causes *Ellis–van Creveld syndrome*. Affected individuals have extra fingers, toes, or both (Section 12.10). The allele might have been rare when a few founders entered Pennsylvania. Now, about 1 in 8 individuals of the community are heterozygous for the allele, and 1 in 200 are homozygous for it.

Genetic drift is the random change in allele frequencies over the generations, brought about by chance alone. The magnitude of its effect is greatest in small populations, such as one that endures a bottleneck.

18.8 Gene Flow

Individuals, and their alleles, move into and away from populations. The physical flow of alleles counters changes introduced by other microevolutionary processes.

Individuals of the same species don't always stay put. A population loses alleles when an individual leaves it for good, an event called *emigration*. The population gains alleles when individuals permanently move in, an event called *immigration*. In both cases, **gene flow** —the physical movement of alleles into and out of a population—occurs. This microevolutionary process counters mutation, natural selection, and genetic drift.

Later chapters will give historical examples of how gene flow has kept separated populations genetically similar. For now, simply consider the acorns that blue jays disperse when they gather nuts for the winter. Each fall, jays visit acorn-bearing oak trees repeatedly, then bury acorns in the soil of home territories that may be as much as a mile away (Figure 18.16). Alleles flowing in with the "immigrant acorns" help decrease genetic differences between stands of oak trees.

Figure 18.16 Blue jay, a mover of acorns that helps keep genes flowing between separate oak populations.

Or think of the millions of people from politically explosive, economically bankrupt countries who seek a more stable home. The scale of their emigrations is unprecedented, but the flow of genes is not. Human history is rich with cases of gene flow that minimized many of the genetic differences among geographically separate groups. Remember Genghis Khan? His genes flowed from China to Vienna (Section 12.10). Similarly, the armies of Alexander the Great brought alleles for green eyes from Greece all the way to India.

Gene flow is the physical movement of alleles into and out of a population, through immigration and emigration. It tends to counter the effects of mutation, natural selection, and genetic drift.

18.9 Adaptation to What? A Word of Caution

LINKS TO
SECTIONS
1.4, 17.3

Observable traits are not always easy to correlate with conditions in an organism's environment.

"Adaptation" is one of those words that have different meanings in different contexts. An individual plant or animal often can quickly adjust its form, function, and behavior. Junipers in inhospitably windy places grow less tall than junipers of the same species in more sheltered places. This is an example of a *short-term* adaptation, because it lasts only as long as the individual plant does.

An **evolutionary adaptation** is some aspect of form, function, behavior, or development that improves the odds for surviving and reproducing in a particular environment. This is an *outcome* of microevolution—natural selection especially—an enhancement of the fit between the individual and prevailing conditions.

SALT-TOLERANT TOMATOES

As an example of long-term adaptation, compare how tomato species handle salty water. Tomatoes evolved in Ecuador, Peru, and the Galápagos Islands. The type sold most often in markets, *Lycopersicum esculentum*, has eight close relatives in the wild. If you mix ten grams of table salt with sixty milliliters of water, then pour it into the soil around *L. esculentum*'s roots, the plant will wilt drastically in less than thirty minutes (Figure 18.17*a*). Even when the soil has only 2,500 parts per million of salt, this species grows poorly.

Yet the Galápagos tomato (*L. cheesmanii*) survives and reproduces in seawater-washed soils. We know that its salt tolerance is a heritable adaptation. How? Crosses of a wild species with the commercial species yield a small, edible F$_1$ hybrid. The hybrid tolerates

Figure 18.17 (**a**) Severe, rapid wilting of one commercial tomato plant (*Lycopersicum esculentum*) that absorbed salty water. (**b**) Galápagos tomato plant, *L. cheesmanii*, which stores most absorbed salts in its leaves, not in its fruits.

irrigation water that is two parts fresh and one part salty. It is getting attention in areas where fresh water is scarce and where salts have built up in croplands.

It may take modification of only a few traits to get new salt-tolerant plants. Revving up just one gene for a sodium–hydrogen ion transporter helps the tomato plants use salty water and still bear edible fruits.

NO POLAR BEARS IN THE DESERT

You can safely bet that a polar bear (*Ursus maritimis*) is finely adapted to the icy Arctic, and that its form and function would be a flop in a desert (Figure 18.18). You

Figure 18.18 Which adaptations of a polar bear (*Ursus maritimus*) won't help in a desert? Which ones help an oryx (*Oryx beisa*)? For each animal, make a tentative list of possible structural and functional adaptations to the environment. Later, after you finish reading Unit VI, see how you can expand the list.

Figure 18.19 Adaptation to what? A heritable trait is an adaptation to specific environmental conditions. Hemoglobin of llamas, which live at high altitudes, has a high oxygen-binding affinity. However, so does hemoglobin of camels, which live at lower elevations.

might be able to make some educated guesses about why that is so. However, detailed knowledge of its anatomy and physiology might make you view it—or any other animal or plant—with respect. How does a polar bear maintain its internal temperature when it sleeps on ice? How can its muscles function in frigid water? How often must it eat? How does it find food? Conversely, how can an oryx walk about all day in the blistering heat of an African desert? How does it get enough water when there is no water to drink? You will find some answers, or at least ideas about how to look for them, in the next three units of this book.

ADAPTATION TO WHAT?

Bear in mind, it is not always easy to identify a direct relationship between adaptation and the environment. For instance, the prevailing environment may be very different from the one in which a trait evolved.

Consider the llama. It is native to the cloud-piercing peaks of the Andes in western South America (Figure 18.19). The llama lives 4,800 meters (16,000 feet) above sea level. Compared to humans at lower elevations, its lungs have more air sacs and blood vessels. The llama heart has larger chambers, so it pumps larger volumes of blood. Llamas do not have to produce extra blood cells, as people do when they move permanently from lowlands to high elevations. (Extra cells make blood "stickier," so the heart has to pump harder.) But the most publicized adaptation is this: Llama hemoglobin is better than ours at latching on to oxygen. It picks up oxygen in the lungs far more efficiently.

Superficially, at least, the oxygen-binding affinity of llama hemoglobin appears to be an adaptation to thin air at high altitudes. Is it? Apparently not.

Llamas are in the same family as dromedary camels. Both share camelid ancestors that evolved in Eocene grasslands and deserts of North America. Later, the ancestors went their separate ways. Forerunners of camels reached Asia's low-elevation grasslands and deserts by a land bridge, which later submerged when the sea level rose. Forerunners of llamas moved down the Isthmus of Panama and on into South America.

Intriguingly, a dromedary camel's hemoglobin also shows a high oxygen-binding capacity. So if the trait arose in a shared ancestor, then how was it adaptive at *low* elevations? We know camels and llamas didn't just *happen* to evolve in the same way. They are close kin, and their most recent ancestors lived in very different environments with different oxygen concentrations.

Who knows why the trait was originally favored? Eocene climates were alternately warm and cool, and hemoglobin's oxygen-binding capacity does go down as temperatures go up. Did it prove adaptive during a long-term shift in climate? Or were its effects neutral at first? What if the allele for efficient hemoglobin was fixed in an ancestral population simply by chance?

Use these "what-ifs" as a reminder to think about observable traits and their presumed connection with a given environment. Identifying the connections takes a great deal of research and experimental tests.

A long-term, heritable adaptation is any aspect of form, function, behavior, or development that contributes to the fit between an individual and its environment.

An adaptive trait improves the odds of surviving and reproducing, or at least it did so under conditions that prevailed when genes for the trait first evolved.

Summary

Section 18.1 Individuals of a population generally have the same number and kinds of genes for the same traits. Alleles are different molecular forms of a gene. Individuals who inherit different allele combinations vary in details of one or more traits. An allele at any locus may become more or less common relative to other kinds or may be lost.

Mutations are rare in individuals, but they have accumulated in natural populations of all lineages. Mutations are the original source of alleles, the raw material for evolution.

Microevolution refers to changes in allele frequencies of a population brought about by mutation, natural selection, genetic drift, and gene flow (Table 18.1).

Section 18.2 Genetic equilibrium is a state in which a population is not evolving. According to the Hardy–Weinberg equilibrium formula, this occurs only if there is no mutation, the population is infinitely large and isolated from all other populations of the species, there is no natural selection, mating is random, and all individuals survive and produce the same number of offspring. Deviations from this theoretical baseline indicate microevolution is in play.

Biology⊘Now
Investigate gene frequencies and genetic equilibrium with the interaction on BiologyNow.

Section 18.3 Natural selection is the outcome of differences in reproduction among individuals of a population that show variations in their shared traits. Three major modes are directional, stabilizing, and disruptive selection. Selection pressures operating on the range of phenotypic variation shift or maintain allele frequencies in the population's gene pool.

Table 18.1	Summary Definitions for Microevolutionary Events
Mutation	A heritable change in DNA; original source of alleles in a population
Natural selection	Outcome of differences in reproduction among individuals of a population that show variation in their shared, heritable traits. Can shift the range of phenotypes in a consistent direction, disrupt it, or stabilize it
Genetic drift	Random changes in a population's allele frequencies through the generations as an outcome of chance alone
Gene flow	Individuals move their alleles into and out of a population by way of immigration and emigration; tends to counter the changes caused by mutation, natural selection, and genetic drift

Section 18.4 Directional selection shifts the range of phenotypic variation in a consistent direction. The individuals at one end of the range of variation are selected against and those at the other end are favored.

Biology⊘Now
View the animation of directional selection on BiologyNow.
Read the InfoTrac article "AIDS in Africa Has Potential to Affect Human Evolution,"AIDS Weekly, June 2001.

Section 18.5 Stabilizing selection works against extremes in the range of phenotypic variation, and it favors intermediate forms. Disruptive selection favors forms at both extremes of the range; individuals in the intermediate range are selected against.

Biology⊘Now
View the animation of disruptive and stabilizing selection on BiologyNow.
Read the InfoTrac article "Portraits of Evolution: Studies of Coloration in Hawaiian Spiders," Geoffrey S. Oxford, Rosemary G. Gillespie, Bioscience, July 2001.

Section 18.6 Sexual selection, by females or males, leads to forms of traits that favor reproductive success. Persistence in phenotypic differences between males and females (sexual dimorphism) is one outcome.

Selection may result in balanced polymorphism, with nonidentical alleles for a trait being maintained over time at relatively high frequencies.

Biology⊘Now
Read the InfoTrac article "High-Risk Defenses," Gregory Cochran, Paul W. Ewald, Natural History, Feb. 1999.

Section 18.7 Genetic drift is a random change in a population's allele frequencies over time due to chance occurrences alone. It tends to lead to the homozygous condition and loss of genetic diversity.

The effect of genetic drift is most pronounced in very small populations, such as ones that have passed through a bottleneck or that arose from a small group of founders. Genetic drift has less effect on inbred populations, which are characterized by nonrandom mating of very close relatives.

Biology⊘Now
Learn more about genetic drift with the interaction on BiologyNow.

Section 18.8 Gene flow moves alleles into or out of a population by immigration or emigration. The process helps keep populations of the same species genetically alike by countering the effects of mutation, natural selection, and genetic drift.

Section 18.9 Long-term, heritable adaptations are aspects of form, function, behavior, or development that improve the chance of surviving and reproducing, or at least did so under conditions that prevailed when genes for the trait first evolved.

Often it is not easy to correlate an adaptive trait with the particular environmental conditions to which it is assumed to be adapted.

Self-Quiz

Answers in Appendix II

1. Individuals don't evolve, _____ do.

2. Biologists define evolution as _____ .
   a. purposeful change in a lineage
   b. heritable change in a line of descent
   c. acquiring traits during the individual's lifetime
   d. both a and b

3. _____ is the original source of new alleles.
   a. Mutation   d. Gene flow
   b. Natural selection   e. All are original sources of
   c. Genetic drift   new alleles

4. Natural selection may occur when there are _____ .
   a. differences in forms of traits
   b. differences in survival and reproduction among
      individuals that differ in one or more traits
   c. both a and b

5. Directional selection _____ .
   a. eliminates common forms of alleles
   b. shifts allele frequencies in a consistent direction
   c. favors intermediate forms of a trait
   d. works against adaptive traits

6. Disruptive selection _____ .
   a. eliminates uncommon forms of alleles
   b. shifts allele frequencies in one direction only
   c. doesn't favor intermediate forms of a trait
   d. both b and c

7. Sexual selection, especially competition between
   males for access to fertile females, frequently influences
   aspects of body form and leads to _____ .
   a. inbreeding   c. sexual dimorphism
   b. genetic drift   d. both b and c

8. The persistence of malaria and sickle-cell anemia
   in a population is a case of _____ .
   a. bottlenecking   c. natural selection
   b. balanced   d. artificial selection
      polymorphism   e. both b and c

9. _____ tends to counter changes that occur in the
   allele frequencies among populations of a species.
   a. Genetic drift   c. Mutation
   b. Gene flow   d. Natural selection

10. Match the evolution concepts.
    ____ gene flow   a. source of new alleles
    ____ natural   b. changes in a population's allele
         selection   frequencies due to chance alone
    ____ mutation   c. allele frequencies change owing to
    ____ genetic   immigration, emigration, or both
         drift   d. outcome of differences in survival,
                           reproduction among individuals
                           of a population that vary in the
                           details of shared traits

Additional questions are available on **Biology ⊜ Now**™

Critical Thinking

1. Occasionally, a few of the families in a remote region of
Kentucky produce *blue offspring*, a condition caused by an
autosomal recessive disorder. Skin of affected individuals
appears dark blue. Homozygous individuals do not have

Figure 18.20 Two designer dogs: the Great Dane (*legs, left*) and
the chihuahua (*possibly fearful of being stepped on, right*).

the enzyme that maintains hemoglobin in its normal
molecular form. Without it, a blue form of hemoglobin
accumulates in blood and shows through the skin.

Formulate a hypothesis to explain the recurrence of
the blue offspring trait among a cluster of families.

2. Martha is studying a population of tropical birds. The
males have brightly colored tail feathers and the females
don't. She suspects this difference is maintained by sexual
selection. Design an experiment to test her hypothesis.

3. About 50,000 years ago, humans began domesticating
wild dogs. By 14,000 years ago, they started to favor new
varieties (breeds) by way of artificial selection. Individual
dogs having desirable forms of traits were selected from
each new litter and, later, encouraged to breed. Those with
undesired forms of traits were passed over.

After favoring the pick of the litter for hundreds or
thousands of generations, we ended up with sheep-herding
border collies, badger-hunting dachshunds, bird-fetching
retrievers, and sled-pulling huskies. And at some point we
began to delight in the odd, extraordinary dog.

In practically no time at all, evolutionarily speaking,
we picked our way through the pool of variant dog alleles
and came up with such extreme breeds as Great Danes and
chihuahuas (Figure 18.20).

Sometimes the canine designs have exceeded the limits
of biological common sense. How long would a tiny, nearly
hairless, nearly defenseless, finicky-eating chihuahua last
in the wild? Not long. What about English bulldogs, bred
for a stubby snout and compressed face? Breeders thought
these traits would let the dogs get a better grip on the nose
of a bull. (Why they wanted dogs to bite bulls is a story in
itself.) So now the roof of the bulldog mouth is ridiculously
wide and often flabby, so bulldogs have trouble breathing.
Sometimes they get so short of breath they pass out.

Why do you suppose many people easily accept that
artificial selection practices can produce startling diversity
but will not accept that natural selection might do the same
in the wild?

Last of the Honeycreepers?

More than 5 million years ago, Kauai rose above the surface of the sea. It was the first of the big islands of the Hawaiian Archipelago. Several million years later, a few quite possibly terrified finches reached it after bobbing 4,000 kilometers (2,500 miles) across the open ocean. Were they unwilling pioneers, blown away from the mainland during a fierce storm? We may never know, but their chance geographic dispersal was the start of something big.

No predatory mammals had preceded the finches onto that isolated, volcanically born island. But tasty insects and plants that bore tender leaves, nectar, seeds, and fruits were already there. The finches thrived. Their descendants quickly radiated into habitats along the coasts, through dry lowland forests, and into rain forests of the highlands.

Between 1.8 million and 400,000 years ago, volcanic eruptions created the rest of the archipelago. Generation after generation, descendants of the first finches traveled on the winds to vacant habitats in the new islands. They foraged in many shrublands and forests, each with special food sources and nesting sites. Diverse agents of natural selection operated in each place, and differences in bill sizes and shapes, feather coloration and patterns, and territorial songs evolved. In this way, a spectacular family of birds, the Hawaiian honeycreepers, originated.

One existing Hawaiian honeycreeper has a bill that fits in the long, curving nectar tubes of *Lobelia* flowers (Figure 19.1). One probes tree bark with its sickle-shaped upper bill, then scoops out beetle larvae with its shovel-shaped lower bill. Other species use a thickened, strong, parrotlike bill to crush or pry open hard seed pods. The po'-ouli (Figure 19.2) is the only species that preferentially eats native tree snails.

Ironically, the very isolation that favored specialized adaptations to conditions in unique habitats made these birds vulnerable to extinction. When conditions changed, they had nowhere to go. They had no built-in defenses against predatory mammals and avian diseases of the mainland, against humans who coveted cloaks made of their eye-catching feathers, or against climate change.

Accompanying humans to the islands were brown tree snakes, rats, cats, and other voracious predators. People also imported chickens and other birds that happened to be infected with disease agents. Over time, people cleared more and more of the forests. Imported crop plants and plant-eating mammals became established. The Hawaiian honeycreeper habitats shrank. Today, with a long-term increase in global temperatures—global warming—the forests at higher elevations are not as cool as they once were. They have been infiltrated by mosquitoes, which

Figure 19.1 *Left*, the Hawaiian honeycreeper known informally as Iiwi (*Vestiaria coccinea*). It evolved in the Hawaiian Archipelago (*right*), far from the mainland. It is a descendant of a spectacular adaptive radiation—one of the patterns explained in this chapter.

Figure 19.2 Male po'ouli—rare, old, and missing one eye. Ecologists captured this small honeycreeper on the east slope of Haleakala, Hawaii, as part of a last-ditch effort to save the remaining population of three birds. This male was already suffering the effects of avian malaria, and it died in 2004. The perpetuation of its species—which was not even known about until 1973—now rests on the two remaining birds. The two have not been seen in many months.

thrive in warm climates. Mosquitoes happen to be vectors for pathogens that cause avian malaria and other diseases.

At one time there were approximately fifty species of Hawaiian honeycreepers. As many as twenty-four species colonized a single island. Half of the known species are extinct. The initial wave of extinction followed the arrival of the first Polynesians, and another ten species are now endangered. The remaining species are being studied in earnest, and efforts are under way to protect them. It may be a case of too little, too late; but time will tell.

How do we know so much about a group of birds on an island chain in the middle of the Pacific Ocean? Scientific theories and tools, particularly radiometric dating and automated gene sequencing, helped shine light on their rise and impending fall. The age of volcanic rocks on each island, as well as the DNA of different species, were among the clues. Such clues give us glimpses into **macroevolution** —the long-term patterns, rates, and trends in the origin and ultimate fate of Earth's many millions of species.

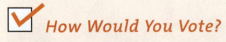

How Would You Vote?

Often, when a species is on the brink of extinction, some individuals are captured and brought to zoos for captive breeding programs. Some people object to this practice. They say keeping a species alive in a zoo is a distraction from more meaningful conservation efforts, and captive animals seldom are successfully restored to the wild. Do you support captive breeding of highly endangered species? See BiologyNow for details, then vote online.

Key Concepts

HOW DO SPECIES ARISE?

All sexually reproducing species consist of one or more populations of individuals that interbreed under natural conditions, produce fertile offspring, and are reproductively isolated from other such populations. Section 19.1

MODELS FOR SPECIATION

Speciation is a process that varies in details and duration among lineages. It starts when gene flow stops between populations of a species. Microevolutionary events occur independently in the reproductively isolated populations. The process ends when daughter species form. Sections 19.2, 19.3

PATTERNS IN THE HISTORY OF LIFE

The timing, rate, and direction of speciation differ among branches of a lineage and between lineages. Adaptive radiations and extinction punctuate the history of life. Section 19.4

CLASSIFICATION SYSTEMS

Patterns in life's history are being identified and interpreted. Taxonomy identifies, names, and then classifies species. Systematics infers evolutionary relationships by analytical methods. Phylogenetic classification systems are efficient tools for retrieving information about the history of life. Section 19.5

PIECING TOGETHER FAMILY TREES

Biologists construct evolutionary tree diagrams that use derived traits to determine branch points. A current tree subsumes six traditionally defined kingdoms into three domains: Bacteria, Archaea, and Eukarya. It reveals how all species interconnect through shared ancestors, some remote, others recent. Sections 19.6–19.9

Links to Earlier Concepts

Before starting this chapter, be sure you understand how gene flow can help keep populations of the same species genetically similar by countering the impact of mutation, natural selection, and genetic drift (Sections 18.1, 18.8). You will be appying your knowledge of changes in chromosome structure and function (12.8 and 15.4). Quickly review how comparisons of morphology (17.7) and of genes and proteins of different lineages (17.9) yield clues to shared ancestry.

Also reflect on the major geologic forces. They have been a factor in the origin of many species, especially on island chains (17.2, 17.6). You also will be taking a closer look at the three-domain system of classification (1.3).

19.1 Reproductive Isolation, Maybe New Species

LINKS TO
SECTIONS
11.6, 18.1, 18.8

Speciation is a macroevolutionary process. It starts when a population becomes reproductively isolated from others of the species and ends when daughter species have formed.

WHAT IS A SPECIES?

Species is a Latin word that means "kind," as in "one kind of plant." This generic definition does not help much when we are trying to figure out whether, say, a population of plants in one place belongs to the same species as a population of plants somewhere else. You may see variations in traits between the populations and variation within them, because plants can inherit diverse combinations of alleles. Also, some individuals may grow in very different environments that cause changes in gene expression (Section 11.6 and Figure 19.3). In other words, we might not be able to identify a biological species on the basis of appearance alone.

Evolutionary biologist Ernst Mayr came up with a **biological species concept**: A species is one or more groups of individuals that interbreed, produce fertile offspring, and are reproductively isolated from other such groups. This definition is reasonable for species that reproduce sexually—which most species do. It does not apply to asexual reproducers, and it cannot be used to interpret the fossil record.

A more recent definition has wider applicability: A **species** is one or more populations of individuals that share at least one structural, functional, or behavioral trait—*the legacy of a common ancestor*—that sets them apart from other species. This definition is based on comparative morphology, biochemistry, and the fossil record. It applies to sexually or asexually reproducing species. Unlike the biological species concept, it does not directly address how a species attains and then maintains its separate identity. That clue, for sexual reproducers at least, is *reproductive isolation*—the end of gene exchanges between populations.

REPRODUCTIVE ISOLATING MECHANISMS

Gene flow, recall, is the movement of alleles into and out of a population (Section 18.8). Speciation begins when gene flow, or the potential for it, ends between natural populations. Once it stops, gene pools start to change and populations undergo **genetic divergence**, because mutation, natural selection, and genetic drift are free to operate independently in each one (Section 18.8). As you will see later, speciation may result from gradual genetic divergence. It also may be completed within a few generations, as commonly occurs among flowering plants.

a

b

Figure 19.3 Morphological differences between plants of the same species (*Sagittaria sagittifolia*) growing (**a**) in water and (**b**) on land. The leaf shapes are responses to different environmental conditions, not to different genetic programs.

Figure 19.4 *Animated!* (**a**) Mechanical isolation. Few pollinating insects fit as well as wasps on a zebra flower. Petals form a landing platform below stamens.

(**b**) Temporal isolation. *Magicicada septendecim*, a periodical cicada that matures underground and emerges to reproduce every seventeen years. Its populations often overlap the habitats of a sibling species (*M. tredecim*), which reproduces every thirteen years. Adults live only a few weeks.

(**c**) Behavioral isolation. Courtship displays precede sex among many kinds of birds, including these albatrosses. Individuals recognize tactile, visual, and acoustical signals, such as a prancing dance followed by back arching, a skyward pointing bill and an exposed throat, and wing spreading.

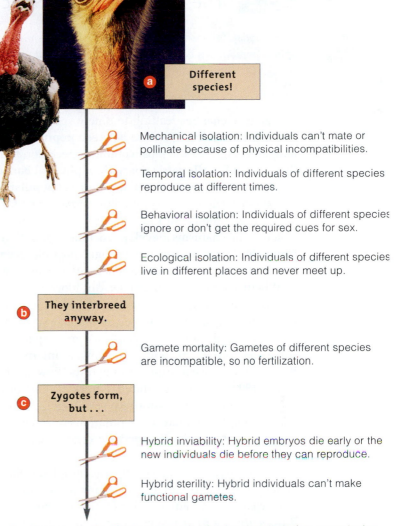

Figure 19.5 *Animated!* When certain reproductive isolating mechanisms prevent interbreeding. There are barriers to (**a**) getting together, mating, or pollination, (**b**) successful fertilization, and (**c**) survival, fitness, or fertility of hybrid embryos or offspring.

a Different species!

Mechanical isolation: Individuals can't mate or pollinate because of physical incompatibilities.

Temporal isolation: Individuals of different species reproduce at different times.

Behavioral isolation: Individuals of different species ignore or don't get the required cues for sex.

Ecological isolation: Individuals of different species live in different places and never meet up.

b They interbreed anyway.

Gamete mortality: Gametes of different species are incompatible, so no fertilization.

c Zygotes form, but . . .

Hybrid inviability: Hybrid embryos die early or the new individuals die before they can reproduce.

Hybrid sterility: Hybrid individuals can't make functional gametes.

No offspring, sterile offspring, or weak offspring that die before reproducing

Either way, **reproductive isolating mechanisms** evolve. All of these heritable aspects of body form, function, or behavior block interbreeding between populations. *Prezygotic* mechanisms, as in Figure 19.4, stop cross-pollination or cross-breeding, the formation of gametes, or fertilization. *Postzygotic* mechanisms kill hybrids or make them weak or infertile. Let us start off with the prezygotic isolating mechanisms listed in Figure 19.5*a,b*.

Mechanical isolation. The body parts of a species are not a physical match with those of a species that could otherwise serve as a mate or pollinator. Figure 19.4*a* shows the fit between a zebra plant and its preferred pollinator. Similarly, the pollen-bearing stamens of the flowers of one sage species extend above petals that act as a landing platform. Big-bodied pollinators get a dusting of pollen when they land and collect nectar. Pollen-gathering bees are not large enough to brush against these stamens. But the other sage species has its stamens poised above a bee-sized platform that is too small and fragile to hold big, heavy pollinators.

Temporal isolation. Diverging populations cannot interbreed when their timing of reproduction differs. Cicada species that differ in form and behavior often live in the same habitat in the eastern United States. They all mature underground and feed on juicy roots. Every 17 years, three species emerge and reproduce (Figure 19.4*b*). Each one has a *sibling* species of similar form and behavior. But siblings emerge on a 13-year cycle. This means that each species and its sibling do not get together except once every 221 years!

Behavioral isolation. Behavioral differences bar gene flow between related species. Before male and female birds copulate, they may engage in courtship displays (Figure 19.4*c*). A female bird is genetically prewired to recognize the singing, wing spreading, prancing, or head bobbing of a male of her species as an overture to sex. Females of different species usually do not.

Ecological isolation. Populations occupying different microenvironments may be ecologically isolated. Two manzanita species live in seasonally dry foothills of the Sierra Nevada, one at elevations between 600 and 1,850 meters, the other between 750 and 3,350 meters. They hybridize rarely, and only where the two ranges overlap. Water-conserving mechanisms operate in dry seasons. But one species is adapted to sites where water stress is not intense. The other lives in drier, exposed sites on rocky hillsides, so cross-pollination is unlikely.

Gamete mortality. Gametes of different species may have molecular incompatibilities. Example: If pollen lands on a plant of another species, it usually does not respond to the plant's molecular signals to germinate.

Postzygotic isolating mechanisms act in an embryo (Figure 19.5*c*). Unsuitable interactions among genes or gene products cause early death, sterility, or weak hybrids with low survival rates. Certain hybrids are sturdy but sterile. Mules, which are the offspring of a female horse and male donkey, are infertile hybrids.

A species is one or more populations of individuals having a unique common ancestor. Its individuals share a gene pool, produce fertile offspring, and remain reproductively isolated from individuals of other species.

Speciation is the process by which daughter species form from a population or subpopulation of a parent species. The process varies in its details and duration, but all modes of speciation are based on reproductive isolation.

19.2 The Main Model for Speciation

LINKS TO
SECTIONS
17.2, 17.6, 17.9

*Three models for speciation differ in their basic premise
of how populations become reproductively isolated.*

START WITH GEOGRAPHIC ISOLATION

The genetic changes leading to a new species usually
begin with *physical separation* between populations, so
allopatry might be the most common speciation route.
By a model for **allopatric speciation**, physical barriers
stop gene flow among populations or subpopulations
of a species. (*Allo–* means different; *patria* can be taken
to mean the homeland.) In both groups, reproductive
isolating mechanisms develop. In time, speciation is
complete. Interbreeding is no longer possible even if
daughter species come into contact with one another.

Whether a geographic barrier can block gene flow
depends on an organism's means of travel (deliberate
or accidental), how fast it can travel, and whether it
is inclined to disperse. Populations of most species are
some distance apart, and gene flow is intermittent.
Barriers may arise abruptly and end the flow entirely.
In the 1800s, a major earthquake buckled part of the
Midwest and the Mississippi River changed course. It
cut through the habitats of populations of insects that
could not swim or fly. It ended the gene flow between
those adjoining populations.

The fossil record suggests that geographic isolation
generally happens slowly. For example, it happened
after vast glaciers advanced into North America and
Europe during the ice ages and cut off populations of
plants and animals from one another. After glaciers
retreated and the descendants of related populations
met, some were no longer reproductively compatible.
They were separate species. Genetic divergence was
not as great between other separated populations, so
descendants still interbred. In their case, reproductive
isolation was incomplete; speciation did not follow.

Also, remember how Earth's crust is fractured into
gigantic plates? Slow, colossal movements inevitably
alter the configurations of land masses (Section 17.6).
As Central America formed, part of an ancient ocean
basin was uplifted, and it became a land bridge—now
called the Isthmus of Panama. Some camelids crossed
the bridge into South America. Geographic separation
led to new species: llamas and vicunas (Figure 19.6).

THE INVITING ARCHIPELAGOS

An **archipelago** is an island chain some distance from
a continent. Many chains are so close to the mainland
that gene flow is more or less unimpeded, so there is
little if any speciation. The Florida Keys are like this.
As you read earlier, the Hawaiian Islands, Galápagos
Islands, and other remote, isolated archipelagos favor
adaptive radiations and speciation (Figures 17.3 and
19.1). The islands are only the tops of volcanoes that
started building up on the seafloor. In time they broke
the surface of the ocean. We can therefore assume that
their fiery surfaces were initially barren, with no life.

In one view, winds or ocean currents carry a few
individuals of some mainland species to such islands,
as shown in Figure 19.7a. Descendants colonize other

Figure 19.6 Allopatric
speciations. The earliest
camelids, no bigger than
a jackrabbit, evolved in the
Eocene grasslands and
deserts of North America.
By the end of the Miocene,
they included the now-extinct
Procamelus. The fossil record
and comparative studies
indicate that this may have
been the common ancestral
stock for llamas (**a**), vicunas
(**b**), and camels (**c**). One of
the descendant lineages
dispersed into Africa and
Asia and evolved into modern
camels. A different lineage,
ancestral to the llamas and
vicunas, dispersed into
South America after gradual
crustal movements formed
a land bridge between the
two continents.

Late Eocene paleomap, before a
land bridge formed between North
and South America. At that time,
North America and Eurasia were
still connected by a land bridge

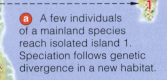 **a** A few individuals of a mainland species reach isolated island 1. Speciation follows genetic divergence in a new habitat.

 **b** Much later, a few individuals of the new species colonize nearby island 2. In this new habitat, speciation follows genetic divergence.

 **c** Speciation may follow colonization of islands 3 and 4. It may follow invasion of island 1 by genetically different descendants of the ancestral species.

Akepa (*Loxops coccineus*) Insects, spiders from buds twisted apart by bill; some nectar; high mountain rain forest	Akekee (*L. caeruleirostris*) Insects, spiders, some nectar, high mountain rain forest	Nihoa finch (*Telespiza ultima*) Insects, buds, seeds, flowers, seabird eggs; rocky or shrubby slopes	Palila (*Loxioides bailleui*) Mamane seeds ripped from pods; buds, flowers, some berries, insects; high mountain dry forests	Maui parrotbill (*Pseudonestor xanthrophrys*) Rips dry branches for insect larvae, pupae, caterpillars; mountain forest with open canopy, dense underbrush	Apapane (*Himatione sanguinea*) Nectar, especially of ohi'a-lehua flowers; caterpillars and other insects; spiders; high mountain forests

Po'ouli (*Melamprosops phaeosoma*) Tree snails, insects in understory; last known male died in 2004	Alauahio (*Paroreomyza montana*) Bark or leaf insects; some nectar, high mountain rain forest	Kauai Amakihi (*Hemignathus kauaiensis*) Bark-picker; insects, spiders; nectar; high mountain rain forest	Akiapolaau (*H. munroi*) Probes, digs insects from big trees; high mountain rain forest	Akohekohe (*Palmeria dolei*) Mostly nectar from flowering trees; some insects, pollen; high mountain rain forest	Iiwi (*Vestiaria coccinea*) Mostly nectar (Ohia tree flowers, lobelias, some mints); some insects, high mountain rain forest

 d The ancestor of Hawaiian honeycreepers might have resembled this housefinch (*Carpodacus*), based on morphological studies, and comparisons of chromosomal DNA and mitochondrial DNA sequences for proteins, such as cytochrome *b*.

Figure 19.7 *Animated!* (**a–c**) Allopatric speciation on an isolated archipelago. (**d**) Twelve of fifty-seven known species and subspecies of Hawaiian honeycreepers, with a sampling of their dietary and habitat preferences. Honeycreeper bills are adapted to diverse foods, such as insects, seeds, fruits, and nectar in floral cups.

islands that form in the chain. Habitats and selection pressures differ within and between these islands, so allopatric speciation proceeds by way of divergences. Later, new species may even invade islands that were colonized by their ancestors. Distances between islands in archipelagos are enough to favor divergence but not enough to stop the occasional colonizers.

The big island of Hawaii formed less than 1 million years ago. Its habitats range from old lava beds, rain forests, and grasslands to snow-capped volcanoes. The first birds to colonize it found a buffet of fruits, seeds, nectars, tasty insects, and few competitors for them. The near absence of competition spurred rapid speciations into vacant adaptive zones. Figure 19.7*d*

shows some of the Hawaiian honeycreepers described earlier. Like thousands of other species of animals and plants, they are unique to this island. As still another example of their potential for speciation, the Hawaiian Islands combined make up less than 2 percent of the world's land masses. Yet they are the original home of 40 percent of all species of fruit flies (*Drosophila*).

By one allopatric speciation model, some type of physical barrier intervenes between populations or subpopulations of a species and prevents gene flow among them. Gene flow ends, and genetic divergences give rise to daughter species.

19.3 Other Speciation Models

LINKS TO
SECTIONS 4.10,
12.8, 15.4, 17.9

There is evidence that some species have arisen and are being maintained by less common mechanisms in which environmental barriers do not play a role.

ISOLATION WITHIN THE HOME RANGE

By the model for **sympatric speciation**, a species may form *within* the home range of an existing species, in the absence of a physical barrier. *Sym–* means together with, as in "together with others in the homeland."

Evidence From Cichlids in Africa   In Cameroon, West Africa, many species of freshwater fishes called cichlids may have arisen by sympatric speciation. The fish live in lakes that formed in the collapsed cones of small volcanoes (Figure 19.8). The cichlids probably colonized the lakes before volcanic action severed the inflow from a nearby river system.

Figure 19.8   A small, isolated crater lake in Cameroon, West Africa, where different species of cichlids may have originated by way of sympatric speciation.

Figure 19.9   Love those polyploids! Among them are several cotton species (including the kind shown here), sugarcane, seedless watermelons, bananas, plums, sweet potatoes, coffee plants with 22, 44, 66, or 88 chromosomes, and marigolds, azaleas, and lilies.

Remember how gene sequences can be compared (Section 17.9)? Ulrich Schliewen looked at differences in nuclear DNA and mitochondrial DNA for eleven cichlid species in Barombi Mbo, one of the small crater lakes. He also compared the samples with DNA from cichlid species in nearby lakes and rivers. He found that cichlid species in Barombi Mbo are more closely related to one another than to neighboring species. He concluded that all of the Barombi Mbo cichlids are descended from the same ancestral species—and that speciation must have occurred *within* this lake.

What could have cause the divergences that led to speciation? The lake is only 2.5 kilometers across, so it is not likely that cichlid populations were separated from one another by any type of barrier. Also, physical and chemical conditions are uniform throughout the lake. Another point: Cichlids are good swimmers, so individuals of different species often meet up.

However, the Barombi Mbo species do show some *ecological* separation. Feeding preferences put species in different places. Some feed in open waters, others at the lake bottom. Yet they all breed close to the lake bottom, in sympatry. Was this small-scale ecological separation enough to promote sexual selection among potential mates? Possibly. Over time, it may have led to reproductive isolation, then speciation.

Polyploidy's Impact   Reproductive isolation might happen within a few generations through **polyploidy**, in which individuals inherit three or more sets of the chromosomes characteristic of their species (Section 12.8). Either a somatic cell fails to divide mitotically after its DNA is duplicated, or nondisjunction occurs at meiosis and results in an unreduced chromosome number in gametes. Offspring usually cannot breed or mate successfully with the parent species, but they may be able to reproduce asexually.

*Auto*polyploids arise by a doubling of the parental chromosome number. This event arises spontaneously in nature but can be in artificially induced in plant breeding laboratories. Breeders expose dividing plant cells to colchicine which, recall, stops microtubular spindles from forming during mitosis (Section 4.10). Without the spindle, duplicated chromosomes do not separate, and cells with the unreduced chromosome number may function as gametes.

*Allo*polyploids originate through (1) spontaneous or induced hybridization between closely related species and (2) doubling of the chromosome number. Figure 19.9 is one example. As genome studies reveal, many stable allopolyploids originated long ago. The kinds produced in the laboratory may or may not prove to

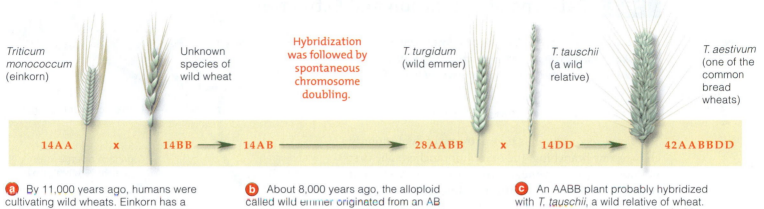

Triticum monococcum (einkorn)   Unknown species of wild wheat   **Hybridization was followed by spontaneous chromosome doubling.**   *T. turgidum* (wild emmer)   *T. tauschii* (a wild relative)   *T. aestivum* (one of the common bread wheats)

14AA   x   14BB → 14AB ⟶ 28AABB   x   14DD ⟶ 42AABBDD

a By 11,000 years ago, humans were cultivating wild wheats. Einkorn has a diploid chromosome number of 14 (two sets of 7). It probably hybridized with another wild wheat species having the same number of chromosomes.

b About 8,000 years ago, the alloploid called wild emmer originated from an AB hybrid wheat plant in which the chromosome number doubled. Wild emmer is tetraploid, or AABB; it has two sets of 14 chromosomes.

c An AABB plant probably hybridized with *T. tauschii*, a wild relative of wheat. Its diploid chromosome number is 14 (two sets of 7 DD). Common bread wheats have a chromosome number of 42 (six sets of 7 AABBDD).

be stable and fertile. Attempts are more successful if the species are close relatives.

Plant speciation is rapid when polyploids produce fertile offspring by self-fertilizing or cross-fertilizing with an identical polyploid. The ancestor of common bread wheat apparently was a wild species, *Triticum monococcum*, which spontaneously hybridized about 11,000 years ago with another wild species (Figure 19.10). Much later in time, a spontaneous chromosome doubling gave rise to *T. turgidum*, an alloploid species with two sets of chromosomes (AABB). Later still, another hybridization resulted in *T. aestivum*, a bread wheat with a chromosome number of 42.

About 95 percent of fern species and 30–70 percent of flowering plants are polyploid species. So are a few conifers, mollusks, insects, and other arthropods, as well as fishes, amphibians, and reptiles. What about mammals? In 1999, A polyploid species of rat with a chromosome number of 102 was found in Argentina.

ISOLATION AT HYBRID ZONES

Parapatric speciation might proceed when different selection pressures operating across a broad region affect populations that are in contact along a common border. Hybrids that form in the contact zone are less fit than individuals on either side of it. Because the hybrids are being selected against, they appear in the hybrid zone only (Figure 19.11).

> By a sympatric speciation model, daughter species arise from a group of individuals within an existing population. Polyploid flowering plants probably formed this way.
>
> By a parapatric speciation model, populations maintaining contact along a common border evolve into distinct species.

Figure 19.10 *Animated!* Presumed sympatric speciation in wheat. Wheat grains 11,000 years old and diploid wild wheats have been found in the Near East, and chromosome analysis indicates that they hybridized. Later, in a self-fertilizing hybrid, homologous chromosomes failed to separate at meiosis, and it produced fertile polyploid offspring. A polyploid descendant hybridized with a wild species. We make bread from grains of their hybrid descendants.

T. barretti
hybrid zone
T. anophthalmus

Figure 19.11 Example of parapatric speciation on the island of Tasmania, directly south of eastern Australia. (**a**) Giant velvet worm, *Tasmanipatus barretti* and (**b**) blind velvet worm, *T. anophthalmus*.

(**c**) Both of these rare species of velvet walking worms live in adjoining regions of northeastern Tasmania. Their habitats overlap in a hybrid zone. Hybrid offspring are sterile, which may be the main reason these two species are maintaining separate identities in the absence of an obvious physical barrier between their habitats.

19.4 Patterns of Speciation and Extinction

LINK TO
SECTION
17.7

*All species, past and present, are related by descent.
They share genetic connections through lineages that
extend back in time to the molecular origin of life.*

BRANCHING AND UNBRANCHED EVOLUTION

The fossil record reveals two patterns of evolutionary change, one branching, the other unbranched. The first is known as **cladogenesis** (from *klados*, branch; and *genesis*, origin). In this pattern, a lineage splits when one or more of its populations become reproductively isolated and diverge genetically. It might be the main speciation pattern. It is the one introduced earlier, in Section 19.1.

In the second pattern, **anagenesis**, changes in allele frequencies and morphology accumulate in a single line of descent. (In this context, *ana*– means renewed.) Directional change is confined within that lineage, as gene flow continues among its populations. In time, allele frequencies and morphology shift so much that the new type differs significantly from the ancestral type, so it is classified as a separate species.

RATES OF CHANGE IN FAMILY TREES

Evolutionary trees summarize information on the relationships among groups. Figure 19.12 can start you thinking about how to construct these tree diagrams. Each branch represents one line of descent from a common ancestor. A *branch point* represents a time of genetic divergence.

When plotted against time, a branch that ends before the present (the treetop) signifies that the lineage is extinct. A dashed line signifies that we know something about the lineage but not exactly where it fits in the tree.

The **gradual model of speciation** holds that species originate by slight morphological changes over long time spans. The model fits with many fossil sequences. For example, sedimentary rock layers often hold vertical sequences of fossilized shells of foraminiferans, as in Figure 19.13. The sequence reflects gradual morphological change.

The **punctuation model of speciation** offers a different explanation for patterns of speciation. Most morphological changes are said to evolve

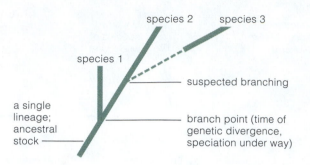

Figure 19.12 Some elements of evolutionary tree diagrams.

in a relatively brief geologic period, within the tens to hundreds of thousands of years when populations are starting to diverge. Directional selection, genetic drift, the founder effect, bottlenecks, or some combination of them favor rapid speciation. The daughter species recover fast from the adaptive wrenching, then they change very little over long periods.

The fossil record shows that stability prevailed for all but 1 percent of the history of most lineages, but it also reveals episodes of abrupt change. As it turns out, both models help explain speciation patterns. Changes have been gradual, abrupt, or both. Species originated at different times and have differed in how long they last. Some did not change much over millions of years; others were the start of adaptive radiations.

ADAPTIVE RADIATIONS

An **adaptive radiation** is a burst of divergences from a single lineage that leads to many new species. This is the pattern that gave rise to the family of Hawaiian honeycreepers. It requires **adaptive zones**, or a set of niches that come to be filled by a group of usually related species. Think of a *niche* as a way of life, such as "burrowing into seafloor sediments" or "catching winged insects in the air at night." Either the lineage enters a vacant adaptive zone or it competes with the resident species well enough to displace them.

You will read more about niches in Chapter 46, in the context of community structure. For now, be aware of two concepts. First, a species must have physical access to a niche when it opens up. Mammals were once distributed in the uniformly tropical regions of Pangea. That supercontinent broke up into huge land

Figure 19.13 Fossilized foraminiferan shells from a vertical sequence of sedimentary rock layers. The first shell (*bottom*) is 64.5 million years old. The most recent (*top*) is 58 million years old. Analysis of shell patterns confirm that the evolutionary order matches the geological sequence.

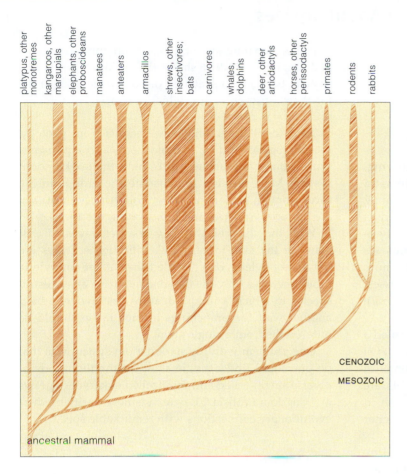

platypus, other monotremes · kangaroos, other marsupials · elephants, other proboscideans · manatees · anteaters · armadillos · shrews, other insectivores; bats · carnivores · whales, dolphins · deer, other artiodactyls · horses, other perissodactyls · primates · rodents · rabbits

CENOZOIC

MESOZOIC

ancestral mammal

Figure 19.14 Adaptive radiation of mammals. Branch widths indicate the range of biodiversity at different times. Mammals arose 220 million years ago but did not start a great radiation until after the K–T impact removed the last of the dinosaurs (page 260). Not all lineages are shown. The 4,000 existing species include shrews, bats, and giant whales.

The photograph shows a fossil of *Eomaia scansoria* (Greek for ancient mother climber). About 125 million years ago, this insectivore crawled on low shrubs and branches. At this writing, it is the earliest placental mammal we know about.

masses, which drifted apart. Habitats and resources changed in different ways, in different places, and set the stage for independent radiations (Figure 19.14).

Second, a species may enter an adaptive zone by a **key innovation**: A chance modification in some body structure or function gives it the opportunity to exploit the environment more efficiently or in a novel way.

Once a species has entered an adaptive zone, genetic divergences can give rise to other species, which can fill a variety of niches within the zone. For example, when the forelimbs of certain vertebrate evolved into wings, novel niches opened up for the ancestors of modern birds and bats (Section 17.7).

EXTINCTIONS—THE END OF THE LINE

An **extinction** is the irrevocable loss of a species. By some estimates, more than 99 percent of all species that ever lived are extinct. The chapter introduction gave examples of typical causes, including imports of new predators and climate change.

In addition to ongoing, small-scale extinctions, the fossil record indicates there were at least twenty or more **mass extinctions**, or catastrophic losses of entire families or other major groups. They differed in size.

For example, 250 million years ago, 95 percent of all known species were abruptly lost. At other times, fewer groups were lost. Afterward, biodiversity slowly recovered as new species filled vacant adaptive zones.

Luck, again, had a lot to do with it. Many species were wiped out by global climate change. When one asteroid struck Earth and the last dinosaurs vanished, mammals were among the survivors that could radiate into vacated adaptive zones. Asteroids, imperceptibly drifting continents, climatic change—all contributed to past patterns of major extinctions and recoveries. In the next unit, you will have plenty of examples.

Lineages have changed gradually, abruptly, or both. Their member species originated at different times and have differed in how long they have persisted.

An adaptive radiation is the rapid origin of many species from a single lineage. It happens when an adaptive zone, a set of similar niches, opens up and the lineage has physical, evolutionary, and ecological access to it.

Repeated and often large extinctions happened in the past. After times of reduced biodiversity, new species originated and occupied new or vacated adaptive zones.

19.5 Organizing Information About Species

So far, you have been thinking about what a species is, how it originates, and what has become of the many, many millions of them that originated. Turn now to what taxonomists do with the information.

SETS OF ORGANISMS—THE HIGHER TAXA

One field of biology, *taxonomy*, deals with identifying, naming, and classifying species. It goes hand in hand with *systematics*, or the study of relationships among organisms. Any organism that has been identified as representing a new species is assigned a unique two-part scientific name, the first part being the genus. As outlined in Chapter 1, species are grouped into more inclusive categories, such as families, orders, classes, phyla (or divisions, which is an equivalent ranking). Figure 19.15 has a few examples.

Each set of organisms in a given category is called a **taxon** (plural, taxa). The sets above the species level are known as the higher taxa, which are the units of classification systems. Most classification systems are now phylogenic, meaning that they reflect perceived evolutionary connections within and between higher taxa as well as patterns of evolutionary change.

A **six-kingdom classification system** promoted by Robert Whittaker prevailed for some time. It assigned all of the prokaryotic species to kingdoms Eubacteria and Archaea, and all single-celled eukaryotes (as well as many multicelled species) to kingdom Protista. It

bestowed separate kingdom status on animals, plants, and fungi. Figure 19.16 shows the six kingdoms of this system. Section 1.3 sketched out a few defining traits for their representatives, which are topics of the next unit of the book.

New fossil finds, and new insights from geology, morphological studies, and biochemical comparisons, caused many researchers to rethink the six-kingdom system. Most decided to subsume the groups into a **three-domain system**, in which the three highest taxa are Bacteria, Archaea, and Eukarya (Figure 19.16).

Why the change? Ongoing research revealed that many of the taxa in earlier classification schemes are not monophyletic, or a "single tribe." A **monophyletic group** includes only the descendants from an ancestral species in which a unique feature first evolved. Said another way, the branchings in each taxon should be outgrowths from a single stem.

One problem with the six-kingdom system was that no one could find a single stem for the thousands of diverse single-celled and multicelled eukaryotic species of "kingdom Protista." Research is now clarifying their evolutionary connections with remarkable speed.

A CLADISTIC APPROACH

Think of each set of species descended from just one ancestral species as a **clade** (from *klados*, a Greek word for branch or twig). A cladistic classification system

KINGDOM	Bacteria	Plantae	Plantae	Animalia	Animalia
PHYLUM	Proteobacteria	Coniferophyta	Anthophyta	Arthropoda	Chordata
CLASS	Epsilonproteobacteria	Coniferopsida	Monocotyledonae	Insecta	Mammalia
ORDER	Campylobacterales	Coniferales	Asparagales	Diptera	Primates
FAMILY	Helicobacteraceae	Cupressaceae	Orchidaceae	Muscidae	Hominidae
GENUS	*Helicobacter*	*Juniperus*	*Vanilla*	*Musca*	*Homo*
SPECIES	*H. felis*	*J. occidentalis*	*V. planifolia*	*M. domestica*	*H. sapiens*
COMMON NAME	none	western juniper	vanilla orchid	housefly	human

Figure 19.15 Taxonomic classification of five species. Each species has been assigned to ever more inclusive sets of organisms—in this case, from species to kingdom.

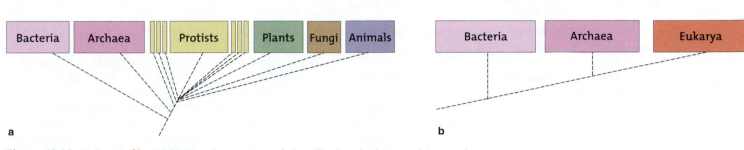

a

b

Figure 19.16 *Animated!* (**a**) Six-kingdom system of classification. In time, protists may be divided into more kingdoms. (**b**) The more recent three-domain system of classification. Protists, plants, fungi, and animals share features that unite them in domain Eukarya.

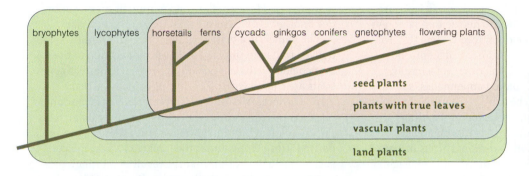

Figure 19.17 Evolutionary tree for land plants, with monophyletic groups nested as sets within sets. All but bryophytes have vascular tissues. All but bryophytes and lycophytes have true leaves. All but bryophytes, lycophytes, horsetails, and ferns produce seeds.

defines clades in terms of the history of divergences, or branch points in time. Only species that share traits derived from the last common ancestor are in the same clade. A **derived trait** is a novel feature that evolved in one species and is present only in its descendants. This emphasis means the descendants within a clade can differ—sometimes exuberantly so—in other traits.

Evolutionary tree diagrams called **cladograms** use the position of branch points from the last shared ancestor to convey inferred evolutionary relationships (phylogenies) among taxa. A cladogram is an estimate of "who came from whom." It has no time bar with absolute dates, so it cannot convey differences in rates of evolution among taxa. Even so, a cladistic approach has already reinforced part of the fossil record, and it is making us reevaluate interpretations of the past.

A tree of life only looks like a simple stick drawing. Many thousands of morphological and biochemical traits were analyzed during the attempts to make its evolutionary connections. For example, as you will see in the next unit of the book, detailed comparisons of genes and ribosomal RNAs have revealed sometimes surprising similarities and differences among groups. The evolutionary tree of life in Section 19.6 is based on such combined evidence.

The derived traits used to construct a cladogram also help us visualize different monophyletic groups as *sets within sets*. For example, Figure 19.17 shows a cladogram for major sets of land plants that have been

nested into ever larger categories. We assume that the cycads, ginkgos, conifers, gnetophytes, and flowering plants form one set, because only they have a common ancestor that was the first seed-producing plant. The seed plants are nested in a larger set—plants with true leaves—that includes horsetails and ferns but excludes lycophytes and bryophytes. Only the bryophytes are not nested in the still-larger set called vascular plants; they do not have tubelike tissues that deliver water and solutes throughout the plant body.

The section to follow shows you how to construct a cladogram. It provides a closer look at the advantages and some of the pitfalls of a cladistic approach.

Taxonomists identify, name, and classify sets of organisms into ever more inclusive categories, the higher taxa.

Classification systems organize and simplify the retrieval of information about species. Phylogenetic systems attempt to reflect evolutionary relationships among species.

Reconstructing the evolutionary history of a given lineage is based on detailed understanding of the fossil record, morphology, life-styles, and habitats of its representatives, and on biochemical comparisons with other groups.

Recent evidence, especially from comparative biochemistry, favors the grouping of organisms into a three-domain system of classification—Archaea, Bacteria, and Eukarya (protists, plants, fungi, and animals).

19.6 How To Construct a Cladogram

LINK TO
SECTION
17.6

In case you would like to know how a cladogram can be constructed, here is a step-by-step approach.

Suppose you want to make a cladogram for vertebrates. You select an *ingroup* of organisms with traits that suggest they might be related—in this case, jaws and paired appendages. You focus on sharks, mammals, crocodiles, and birds because they also differ clearly in some morphological, physiological, and behavioral traits, or characters. Now you must select a different vertebrate that can be used as a reference point for estimating evolutionary distances within the ingroup.

To keep things simple, you check for the presence (+) or absence (–) of seven traits and tabulate them, as in Figure 19.18a. After scanning Chapter 26, you decide that lampreys are only distantly related to the other four vertebrates. For instance, although a tubular structure called a notochord forms in its embryos, as it does for all other vertebrates, only lampreys have no jaws or paired appendages, such as lateral fins and legs. They can be the *outgroup*, the one with the fewest derived traits when compared to the others.

Derived traits, recall, are evidence of morphological divergence and branching in an evolutionary tree. In Figure 19.18b, each zero (0) across the columns of traits for each vertebrate indicates an ancestral condition. Each numeral one (1) means the vertebrate shows the derived trait.

Now you look for derived traits that the selected groups do or do not share. For example, the crocodile, mammal, and bird share five derived traits, but the bird and the shark share only three. You can now make a simple cladogram, although systematists often use many traits of many taxa. Typically they use a computer to analyze data and find the pattern that is best supported by a lot of information.

Figure 19.18c–g shows how a cladogram develops as you keep adding information to it. Start with the presence or absence of jaws and paired appendages, two traits that all

groups except lampreys derived from a common ancestor. What about lungs? Like lampreys, sharks do not have them, so you have identified another branch point in vertebrate evolution. Past that branch point, only mammals have hair, only crocodiles and birds have some form of gizzard. Only birds and their immediate ancestors have feathers.

How do you "read" the final cladogram? Remember, it is an estimate of *relative* relatedness, which implies common ancestry. Birds are more closely related to crocodiles than they are to mammals. Crocodiles are not the ancestor of birds (they are modern organisms, too), but both share a more recent common ancestor than either does with mammals. Birds, crocodiles, and mammals are closer to one another evolutionarily than they are to the shark.

The higher up a branch point is on a family tree, the more derived traits are shared. The lower the position of the branch point between two groups in the diagram, the fewer traits are shared with other groups being investigated.

A few words of caution: Interpretations of evolutionary relationships are more reliable when many traits are used, and there must be strong evidence that shared traits are derived. This helps counter the impact of a bad choice, such as including a trait that is a result of morphological convergence rather than divergence (Section 17.6).

The choice of derived traits is essential. If you were to select body size, for instance, you might wrongly perceive an evolutionary connection between *Sauroposeidon* (a dinosaur that weighed 60 tons), blue whales (mammals that weigh 200,000 pounds), quaking aspen (one plant has 50,000 stems and weighs an estimated 13 million pounds), and a honey mushroom that has been growing for 2,400 years (its underground body extends through 2,200 acres).

Cladograms are only as good as the choices made for their construction—and good choices start with a broad, deep knowledge of life.

Figure 19.18 (a) Charting out a selection of traits among vertebrate groups that can be used for the construction of a simple cladogram.

(b) A trait's absence in an outgroup or ingroup indicates an ancestral state (here indicated by a zero). Its presence in the set of vertebrates selected as the ingroup is taken to mean it is a derived trait, indicated by a numeral one.

(c–g) Step-by-step construction of a cladogram, as explained in the text.

a

Taxon	Notochord in Embryo	Jaws	Paired Appendages	Lungs	Hair	Gizzard	Feathers
Lamprey	+	–	–	–	–	–	–
Shark	+	+	+	–	–	–	–
Crocodile	+	+	+	+	–	+	–
Mammal	+	+	+	+	+	–	–
Bird	+	+	+	+	–	+	+

b

Taxon	Notochord in Embryo	Jaws	Paired Appendages	Lungs	Hair	Gizzard	Feathers
Lamprey	1	0	0	0	0	0	0
Shark	1	1	1	0	0	0	0
Crocodile	1	1	1	1	0	1	0
Mammal	1	1	1	1	1	0	0
Bird	1	1	1	1	0	1	1

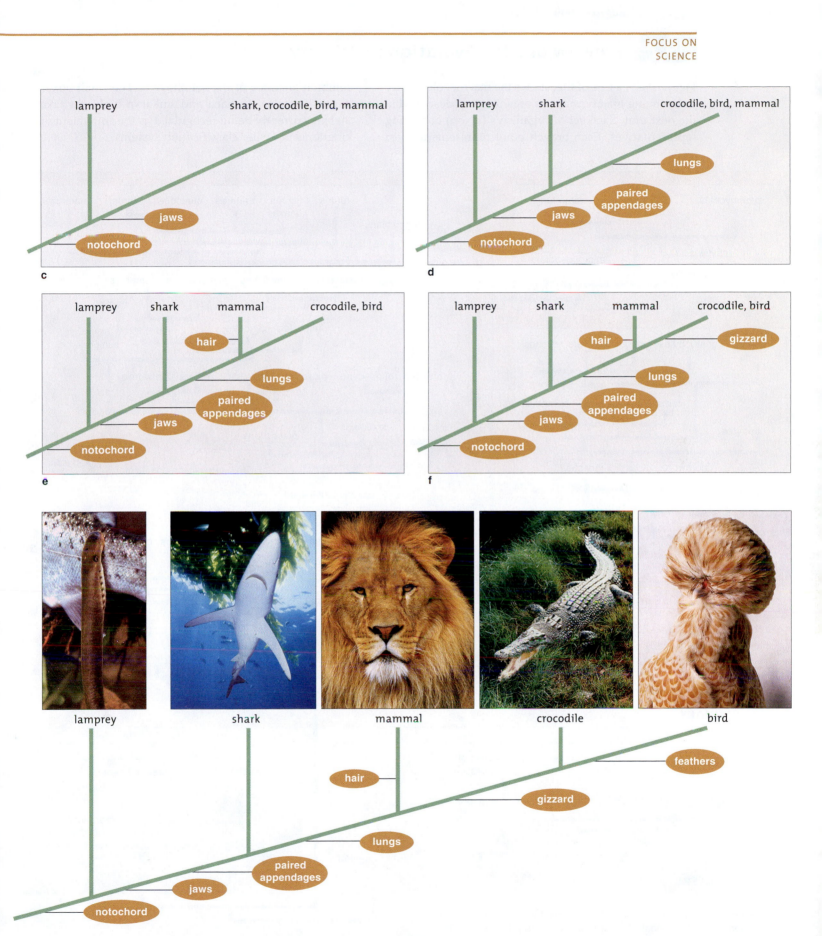

c

lamprey   shark, crocodile, bird, mammal

jaws

notochord

d

lamprey   shark   crocodile, bird, mammal

lungs

paired appendages

jaws

notochord

e

lamprey   shark   mammal   crocodile, bird

hair

lungs

paired appendages

jaws

notochord

f

lamprey   shark   mammal   crocodile, bird

hair   gizzard

lungs

paired appendages

jaws

notochord

lamprey   shark   mammal   crocodile   bird

feathers

hair

gizzard

lungs

paired appendages

jaws

notochord

g The completed diagram, livened up with photographs of representative species.

19.7 Preview of Life's Evolutionary History

Figure 19.19, a tree of life, shows the macroevolutionary links among major groups of organisms, as described in the next unit. Each set of organisms (taxon) has living representatives. Each branch point represents the last common ancestor of the set above it. The small boxes within domains Archaea and Eukarya highlight taxa that are currently being recognized as the equivalent of kingdoms in earlier classification systems.

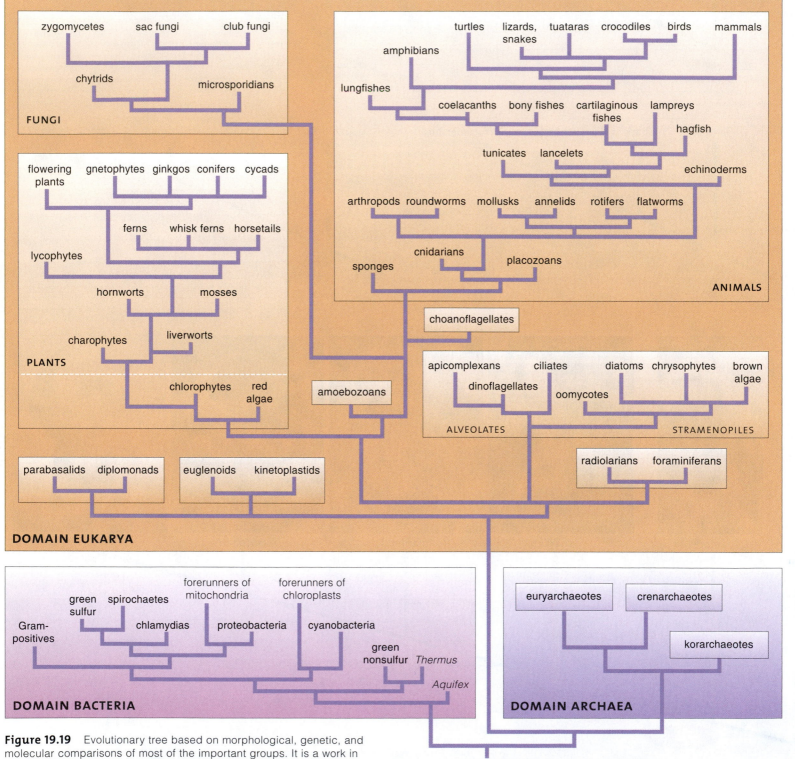

Figure 19.19 Evolutionary tree based on morphological, genetic, and molecular comparisons of most of the important groups. It is a work in progress, subject to refinements as more information comes in.

biochemical and molecular origin of life

19.8 Madeleine's Limbs

So what does macroevolution have to do with you and me? Everything.

In August of 1994, about 900 million years after the first animals appeared on Earth, Madeleine made *her* entrance. Her grandmothers and aunts made a quick count—arms, legs, ears, and eyes, two of each; fully formed mouth and nose—just to be sure these were present and accounted for. One grandmother, having been too long in the company of biologists, had an epiphany as she witnessed Madeleine's birth. In that instant she sensed ancestral connections between the distant past and, through this child, the future.

Madeleine's body plan did not emerge out of thin air. Thirty-five thousand years ago, people just like us were having children just like Madeleine. If we are reading the fossil record correctly, then five million years ago, the offspring of individuals on the road to modern humans resembled her in some respects but not others. Sixty million years ago, primate ancestors of those individuals were giving birth precariously, up in the trees. Two hundred fifty million years ago, the mammalian ancestors of those primates were giving birth—and so on back in time to the very first animals, which had no limbs or eyes or noses at all.

We know little about the very first animals. Yet one thing is clear. By the dawn of the Cambrian, they had given rise to all major groups of invertebrates and to Madeleine's backboned, jawed ancestors. We know this from fossils. For example, one Cambrian community flourished 530 million years ago in, on, and above the dimly lit mud in a submerged basin between a steep reef and the coast of an early continent. About 500 feet below the surface, the water was oxygenated and clear (Figure 19.20). Like a castle built from wet sand, their home was unstable and an underwater avalanche buried them. Over great time spans, compaction and chemical change transformed those small, flattened animals into fossils. By 1909, tectonic activity had moved the fossils high into the eastern mountains of British Columbia, and there a fossil hunter found them.

In the next unit, you will compare body plans of diverse organisms. Such comparisons give insight into evolutionary relatedness and help us construct family trees. As you poke through the tree branches, make use of the evolutionary perspective. At each branch point, the processes of microevolution gave rise to workable changes in body plans. Your collection of conserved and modified traits, and Madeleine's, evolved earlier in countless generations of vertebrates and, even before them, in ancient invertebrate forms.

Our family tree is a record of conserved and derived traits.

Figure 19.20 *Left,* reconstruction of a few Cambrian animals known from fossils of the Burgess Shale in British Columbia. *Right,* Madeleine.

Summary

Section 19.1 Populations of each species share at least one unique trait, a legacy of a common ancestor. In sexually reproducing species, individuals interbreed, produce fertile offspring under natural conditions, and are reproductively isolated from all other species.

If gene flow ends between populations, divergences may lead to new species. Mutation, natural selection, and genetic drift operate independently and may give rise to reproductive isolating mechanisms (Table 19.1). In some cases, reproductive isolation occurs in a few generations.

Prezygotic isolating mechanisms stop interbreeding. They include incompatibilities between reproductive parts or between gametes, differences in reproductive timing or behavior, and ecological restriction to different microenvironments in the same area. The postzygotic mechanisms lead to early death, sterility, or unfit hybrid offspring. They come into play after fertilization.

Biology Now
Use the animation and interaction on BiologyNow to explore how species become reproductively isolated.
Read the Infotrac article "Tracking the Red-Eyed, Sluggish, and Ear-splitting," Tabitha M. Powledge, American Scientist, July 2004.

Section 19.2 By the allopatric speciation model, a geographic barrier cuts off gene flow between two or more populations. Genetic divergence and reproductive isolation are favored and may result in a new species.

Biology Now
Learn more about speciation on an archipelago with the animation on BiologyNow.

Section 19.3 By a sympatric speciation model, populations in physical contact diverge from each other. Polyploid species of many plants and some animals have originated by chromosome doublings and hybridizations.

By a parapatric speciation model, different selection pressures across a broad region act on populations that are in contact along a common border. Unfit hybrids form in the contact zone, so populations on either side diverge independently from each other.

Biology Now
Explore the effects of sympatric speciation in wheat with the animation on BiologyNow.

Section 19.4 Macroevolution refers to the timing, duration, and direction of speciation in the history of life. These features differ among lineages. The major speciation patterns are unbranched (evolution within a single lineage) or branching (divergences from ancestral stock). Most lineages remain stable for long periods, but abrupt episodes of change also have occurred.

Radiations occur in adaptive zones, a similar set of niches that come to be filled by a (usually) related group of species. A niche is a way of life, such as catching insects in the air at night. Species must have physical, evolutionary, and ecological access to these zones.

A key innovation is a chance modification in some body structure or function that lets an organism exploit the environment more efficiently or in a novel way.

Most species are now extinct. Mass extinctions, slow recoveries, and adaptive radiations are major patterns.

Section 19.5 Each species has a unique, two-part scientific name. Taxonomy deals with identifying, naming, and classifying species. Systematics deals with reconstructing life's evolutionary history (phylogeny). In classification systems, sets of organisms (taxa) are organized into ever more inclusive categories as a way to retrieve information about species.

A current three-domain classification system is based largely on phylogenetic evidence. It recognizes three domains: Bacteria, Archaea, and Eukarya. The Eukarya includes diverse lineages known informally as protists, as well as plants, fungi, and animals.

Biology Now
Review biological classification systems with the animation on BiologyNow.

Section 19.6 A cladistic classification system recognizes monophyletic groups. Each group is a clade, a set of species that includes only descendants that display a derived trait, inherited from an ancestor in which that trait first evolved. It is the equivalent of all branches growing from the same point on a stem.

Biology Now
Read the InfoTrac article "How Taxonomy Helps Us Make Sense Out of the Natural World," Sue Hubbell, Smithsonian, May 1996.

Sections 19.7, 19.8 Representing life's history as a tree with branchings from ancestral stems brings clarity to the view that all organisms are related by descent.

Table 19.1	Summary of Processes and Patterns of Evolution	
Microevolutionary Processes		
Mutation	Original source of alleles	Stability or change in heritable traits that define populations, and the species, is the outcome of balances or imbalances among all of these processes. Population size and prevailing conditions in the environment influence the outcome.
Gene flow	Preserves species cohesion	
Genetic drift	Erodes species cohesion	
Natural selection	Preserves or erodes species cohesion, depending on environmental pressures	
Macroevolutionary Processes		
Genetic persistence	The basis of the unity of life. The biochemical and molecular basis of inheritance extends from the origin of first cells through all subsequent lines of descent.	
Genetic divergence	Basis of life's diversity, as brought about by adaptive shifts, branching, and radiations. Rates and times of change varied within and between lineages.	
Genetic disconnect	End of the line for a species. Mass extinctions are catastrophic events in which major groups abruptly and simultaneously are lost.	

Self-Quiz

1. _____ can isolate one population from others.
   a. Structural traits   c. Behavioral traits
   b. Functional traits   d. all of the above

2. Reproductive isolating mechanisms _____ .
   a. stop interbreeding   c. reinforce genetic divergence
   b. stop gene flow   d. all of the above

3. Most species originate by a (an) _____ route.
   a. allopatric   c. parapatric
   b. sympatric   d. parametric

4. In evolutionary trees, a branch point represents a
   ___ ; and a branch that ends represents _____ .
   a. single species; incomplete data on lineage
   b. single species; extinction
   c. time of divergence; extinction
   d. time of divergence; speciation complete

5. Fossil evidence supports the _____ model of
   evolutionary change.
   a. punctuation   b. gradual   c. both are correct

6. *Pinus banksiana, Pinus strobus,* and *Pinus radiata*
   are _____ .
   a. three families of pine trees
   b. three different names for the same organism
   c. three species grouped in the same genus
   d. both a and c

7. Individuals of a monophyletic group _____ .
   a. are all descended from an ancestral species
   b. demonstrate morphological convergence
   c. have a derived trait that first evolved in their
      last shared ancestor
   d. both a and d

8. A(n) _____ classification system reflects presumed
   evolutionary relationships.
   a. epigenetic   c. phylogenetic
   b. tectonic   d. both b and c

9. In modern classification systems, groupings of sets
   of taxa range from _____ to _____ .
   a. kingdom; genera and species
   b. kingdom; genera and domain
   c. genera; domain and kingdom
   d. species; kingdom and domain

10. Match these terms suitably.
    ____ phylogeny   a. now the most inclusive taxon
    ____ extinction   b. tree of branching lineages
    ____ domain   c. many lineages diverge from
    ____ derived trait   one in a new adaptive zone
    ____ cladogram   d. end of a species or lineage
    ____ adaptive   e. evolutionary history of species
         radiation   f. only in descendants of ancestor
                            in which it first evolved

Additional questions are available on Biology ⊛ Now™

Critical Thinking

1. You notice several duck species in the same lake habitat, with no physical barriers hampering the ducks' movements. All the females of the various species look quite similar to one another. But the males differ in the patterning and coloration of their feathers. Speculate on which forms of reproductive isolation may be keeping each species distinct. How does the appearance of the male ducks provide a clue to the answer?

Figure 19.21
Rama the cama displaying his unexpected short temper.

2. *Rama the cama*, a llama-camel hybrid, was born in 1997 (Figure 19.21). Camels and llamas have a shared ancestor but have been separated for 30 million years. Veterinarians collected semen from a male camel that weighed close to 1,000 pounds, then used it to artificially inseminate a female llama one-sixth his weight. The idea was to breed an animal having a camel's strength and endurance and a llama's gentle disposition.

Instead of being large, strong, and sweet, Rama is smaller than expected and has a camel's short temper. Rama resembles both parents, with a camel's long tail and short ears but no hump, and llama-like hooves rather than camel footpads. Now old enough to mate, he is too short to get together with a female camel and too heavy to mount a female llama. He has his eye on Kamilah, a female cama born in early 2002, but will have to wait several years for her to mature. The question is, will any offspring from such a match be fertile?

What does Rama's story tell you about the genetic changes required for irreversible reproductive isolation in nature? Explain why a biologist might not view Rama as evidence that llamas and camels are the same species.

3. Speculate on what might have been a key innovation in human evolution. Describe how that innovation might be the basis of an adaptive radiation in environments of the distant future.

4. Shannon thinks there are too many major taxa and sees no reason to make a new one for something as tiny as archaeans. "Keep them with the other prokaryotes!" she says. Taxonomists would call her a "lumper." But Andrew is a "splitter." He sees no reason to withhold separate status from archaeans simply because they are part of a microscopic world that not many people know about. Which may be the most useful: more or fewer boundaries between groups? Explain your answer.

5. Richard Lenski uses bacterial populations in culture tubes to develop model systems for studying evolution. Bacteria produce several generations in a day. Researchers can store them in the deep freeze, then bring them back to active form, unaltered, to directly compare ancestors and their descendants. Are bacterial models relevant to any evolutionary studies of sexually reproducing organisms? Before you answer, read a short article by P. Raine and M. Travisano entitled "Adaptive Radiation in a Heterogeneous Environment" (*Nature*, July 2, 1998: 69–72).

Looking for Life in All the Odd Places

In the 1960s, microbiologist Thomas Brock was looking for signs of life in the hot springs and pools in Yellowstone National Park (Figure 20.1). He found a simple ecosystem of microscopically small cells, including *Thermus aquaticus.* This prokaryote uses simple carbon compounds dissolved in the water as its energy source. It is known as one of the thermophiles, or "heat lovers," for good reason. *T. aquaticus* withstands temperatures on the order of 80°C (176°F)!

Brock's work had two unexpected results. First, it put researchers on paths that led them to a great domain of life, the Archaea. Second, it led to a faster way to copy DNA and end up with useful amounts of it. *T. aquaticus* happens to make a heat-resistant enzyme, and it can catalyze the polymerase chain reaction—PCR. Synthetic forms of the enzyme helped trigger a revolution in biotechnology.

Bioprospecting became the new game in town. Many companies started to look closely at thermal pools and other extreme environments for species that might yield valuable products. They found forms of life adapted to extraordinary levels of temperature, acidity, alkalinity, salinity, and pressure.

To extreme thermophiles on the seafloor, Yellowstone's hot water would be too cool. They live in the superheated, mineral-rich water near hydrothermal vents. One kind even grows and reproduces at 121°C (249°F). Different species live in acidic springs, where the pH approaches zero, and in highly alkaline soda lakes. In Earth's polar regions, some types cling to life in salt ponds that never freeze and in glacial ice that never melts.

Extreme environments also support some eukaryotic species of ancient lineages. Populations of snow algae tint mountain glaciers red. Another red alga, *Cyanidium caldarium,* is a resident of acidic hot springs. Free-living photosynthetic cells called diatoms live in extremely salty lakes, where the hypertonicity would make cells of most organisms shrivel and die.

What could top that? Nanobes. Australian researchers found nanobes growing 3.8 kilometers (3 miles) below Earth's surface in truly hot rocks—170°C (338°F). Being one-tenth the size of most bacteria, nanobes cannot be observed without electron microscopes. Outwardly, they look something like the simplest fungi (Figure 20.2).

Nanobes are probably too small to be alive. They do not seem to be big enough to hold all of the metabolic machinery that now runs life processes. Even so, nanobes do contain DNA. And they appear to grow. Are they like proto-cells, which preceded the origin of the first living cells? Maybe.

Watch the video online!

Figure 20.1 From a thermal pool in Yellowstone National Park, cells of *Thermus aquaticus,* a prokaryotic species that is immensely admired by recombinant DNA researchers for its heat-resistant enzymes.

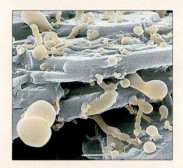

Figure 20.2 Nanobes, possibly like proto-cells. Australian researchers found them in hot rocks far beneath Earth's surface. They are only fifteen to twenty nanometers across; this image has been magnified 20,000 times. However, they do have DNA and other organic compounds enclosed within a membrane, and they grow.

What is the point of these examples? Simply this: *Life can take hold in almost any environment that has sources of carbon and energy.*

This chapter is your introduction to a sweeping slice through time, one that cuts back to Earth's formation and to life's chemical origins. The picture it paints sets the stage for the next unit, which will take you along lines of descent that led to the present range of biodiversity.

The picture is incomplete. Even so, evidence from many avenues of research points to a concept that can help us organize information about an immense journey: *Life is a magnificent continuation of the physical and chemical evolution of the universe, and of the planet Earth.*

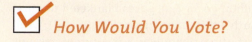

How Would You Vote?

Private companies make millions of dollars selling an enzyme first isolated from cells in Yellowstone National Park. Should the federal government let private companies bioprospect within the boundaries of national parks, as long as it shares in the profits from any discoveries? See BiologyNow for details, then vote online.

Key Concepts

ABIOTIC SYNTHESIS OF ORGANIC COMPOUNDS

The origin and early evolution of life correlate with the physical and chemical evolution of the universe, the stars, and Earth. The first step toward life was the spontaneous formation of complex organic compounds from simpler substances present on the early Earth.   Section 20.1

ORIGIN AND EARLY EVOLUTION OF CELLS

Laboratory studies and computer simulations yield indirect evidence that self-assembly of membranes, combined with chemical and molecular evolution, gave rise to the structural and functional forerunners of cells.

The first cells were anaerobic prokaryotes. Some gave rise to bacteria, others to archaeans and to the ancestors of eukaryotic cells. Evolution of the noncyclic pathway of photosynthesis added oxygen to the atmosphere, which became a major selection pressure.   Sections 20.2, 20.3

HOW THE FIRST EUKARYOTIC CELLS EVOLVED

Organelles help define eukaryotic cells. The nucleus and ER membranes may have evolved through infoldings of the plasma membrane. Mitochondria and chloroplasts may be descended from bacterial parasites or prey that took up permanent residence in host cells.   Section 20.4

VISUAL PREVIEW OF THE HISTORY OF LIFE

A timeline for milestones in the history of life highlights the shared connections among all organisms.   Section 20.5

Links to Earlier Concepts

This chapter starts your survey of the sweep of biodiversity, as introduced in Section 1.3. This is where all of those details of cell metabolism, genetics, and evolutionary theory start to converge and help you make sense of life's fabulous journey. Now you can correlate prokaryotes (4.3) and eukaryotes (4.4) with a timeline of Earth history (17.5).

You will use your knowledge of how organic compounds are assembled (3.2), and of amino acids (3.5), membranes (5.1), enzymes (6.3), and the link between photosynthesis and aerobic respiration (Chapter 7). You may find yourself referring to the sections on DNA replication (13.3), RNAs and protein synthesis (14.1), and the genetic code (14.2). You will consider how the nucleus, ER, mitochondria, and chloroplasts (4.5–4.8) may have originated.

20.1 In the Beginning . . .

LINKS TO
SECTIONS
3.2, 3.5

Life originated when Earth was a thin-crusted inferno, so we may never find evidence of the first cells. Still, answers to three questions can yield clues to their origins. What were conditions like? Did cells emerge as a result of chemical and molecular evolution? Can experimental tests disprove that they did? Let's take a look.

Some clear evening, look up at the moon. *Five billion trillion times* the distance between it and you are the systems of stars, or galaxies, at the edge of the known universe. Light energy travels far faster than anything else, millions of meters a second, yet wavelengths of light that originated from faraway galaxies billions of years ago are just now reaching Earth. By all known measures, all near and distant galaxies in the space of the universe are moving away from one another. The entire universe, it seems, is expanding. One theory of how the colossal expansion started might account for every bit of matter in every living thing.

Think about how you can rewind a videotape on a VCR, then imagine "rewinding" the universe. As you do, the galaxies start moving closer together. After 12 to 15 billion years of rewinding, all galaxies, all matter and space are compressed into a hot, dense volume at one single point. You have arrived at time zero.

That incredibly hot, dense state lasted only for an instant. What happened next is called the **big bang**, the nearly instantaneous distribution of all matter and energy throughout the universe. Within minutes, the temperature dropped a billion degrees. Nuclear fusion reactions created most of the simplest elements, such as helium, which still are the most abundant kinds in the universe. Radio telescopes have detected a relic of the big bang—cooled, diluted background radiation left over from the beginning of time.

Over the next billion years, uncountable numbers of gaseous particles collided, and gravitational forces condensed them into the first stars. When stars were massive enough, nuclear reactions ignited inside them and gave off tremendous light and heat as the heavier elements formed. Stars have a life history, from birth to an often explosive death. In what might be called the original stardust memories, the heavier elements released from dying stars were swept up when new stars formed and helped form even heavier elements.

When explosions of dying stars ripped through our galaxy, they left behind a dense cloud of dust and gas that extended trillions of kilometers in space. As the cloud cooled, countless bits of matter gravitated toward one another. By 5 billion years ago, the shining star of our solar system—the sun—was born.

CONDITIONS ON THE EARLY EARTH

Figure 20.3 shows part of one of the vast clouds in the universe. It is mostly hydrogen gas, along with water, iron, silicates, hydrogen cyanide, ammonia, methane, formaldehyde, and other small inorganic and organic substances. Between 4.6 billion and 4.5 billion years ago, the cloud that became our solar system probably had a similar composition. Clumps of minerals and ice at the cloud's perimeter grew more massive. They became planets; one was the early Earth.

By four billion years ago, gases blanketed the first patches of Earth's thin, fiery crust (Figure 20.4). Most likely, this first atmosphere was a mixture of gaseous hydrogen, nitrogen, carbon monoxide, carbon dioxide. There was little free oxygen. How can we tell? When free oxygen is present, some binds to iron in rocks. However, geologists have discovered that such "rust" did not form until fairly recently in Earth's history.

Figure 20.3 Part of the Eagle nebula, a hotbed of star formation. Each pillar is wider than our solar system. New stars shine on the tips of gaseous streamers.

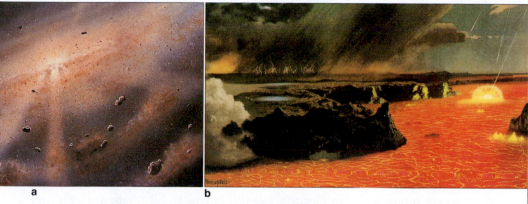
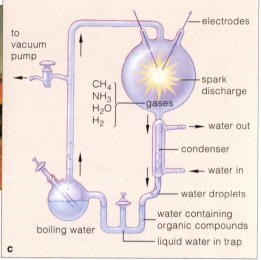

a   b

Figure 20.4 *Animated!* (**a**) What the cloud of dust, gases, rocks, and ice around the early sun might have looked like. (**b**) Less than 500,000 million years later, Earth was a thin-crusted inferno. (**c**) Sketch of the apparatus Stanley Miller used to test whether small organic compounds could form spontaneously in such a harsh environment.

c

The relatively low oxygen levels on the early Earth probably made the origin of life possible. Free oxygen is highly reactive. If it had been present, the organic compounds characteristic of life would not have been able to form and persist. Oxygen radicals would have attacked and destroyed compounds as they formed.

What about water? All of the water that fell on the molten surface would have evaporated at once. After the crust cooled and became solid, however, rainfall and runoff eroded mineral salts from rocks. Over many millions of years, salty water collected in crustal depressions and formed early seas. If liquid water had not accumulated, membranes could not have formed, because they take on their bilayer structure in water. No membrane, no cell, and no life.

ABIOTIC SYNTHESIS OF ORGANIC COMPOUNDS

Cells appeared less than 200 million years after the crust solidified, so complex carbohydrates and lipids, proteins, and nucleic acids must have formed by then. We know that meteorites, Mars, and Earth all formed at the same time, from the same cosmic cloud. Their rocks contain simple sugars, fatty acids, amino acids, and nucleotides, so we can expect that the precursors of biological molecules were on the early Earth, too.

Synthesizing organic compounds requires energy. On the early Earth, lightning, sunlight, or heat from hydrothermal vents might have fueled the reactions. Stanley Miller was the first to test the hypothesis that the simple compounds that now serve as the building blocks of life can form by chemical processes. He put water, methane, hydrogen, and ammonia in a reaction chamber. He kept circulating the mixture and zapping it with sparks to simulate lightning (Figure 20.4c). In

less than a week, amino acids and other small organic compounds had formed in the chemical brew.

Recent geologic evidence suggests that Earth's early atmosphere was not quite like the Miller mixture. But in simulations that used other gases, different organic compounds formed—including certain types that can act as nucleotide precursors of nucleic acids.

By another hypothesis, simple organic compounds formed in outer space. Researchers detect amino acids in interstellar clouds and in some of the carbon-rich meteorites that have landed on Earth. One meteorite found in Australia contains eight amino acids that are identical with those in living organisms.

What about proteins, DNA, and the other *complex* organic compounds? Where could they form? In open water, hydrolysis reactions would have broken them apart as fast as they assembled. By one hypothesis, the clay of tidal flats bound and protected the newly forming polymers. Certain clays contain mineral ions that attract amino acids or nucleotides. Experiments show that once some of these molecules stick to clay, other molecules bond to them and form chains that resemble the proteins or nucleic acids in living cells.

Another hypothesis that is currently getting a lot of attention is this: The first biological molecules were synthesized near hydrothermal vents. Certainly the ancient seafloor was oxygen-poor. Experiments show that amino acids, at least, will condense into protein-like structures when heated in water.

Experiments provide indirect evidence that the complex organic molecules characteristic of life could have formed under conditions that probably prevailed on the early Earth.

20.2 How Did Cells Emerge?

LINKS TO
SECTIONS 5.1,
6.3, 13.4, 14.1, 14.3

Metabolism and reproduction are defining characteristics of life. In the first 600 million years or so of Earth history, enzymes, ATP, and other essential organic compounds assembled spontaneously. If they did so in the same places, their close association might have promoted the start of metabolic pathways and self-replicating systems.

ORIGIN OF AGENTS OF METABOLISM

Before cells appeared, chemical processes may have favored the formation of proteins and other complex organic compounds (Figure 20.5). However proteins originated, their molecular structure dictated their behavior. If some promoted reactions by acting like weak enzymes, they could interact with more amino acids and enzyme helpers, such as metal ions.

Visualize an early estuary, where seawater mixed with mineral-rich water that drained from the land. Beneath the sun's rays, organic molecules got stuck to clay in the mud (Figure 20.6a). At first, there were quantities of an amino acid; call it **D**. Molecules of **D** became incorporated into proteins—until **D** started to run out. Close by, however, was a weakly catalytic protein. This protein could speed the formation of **D** from a plentiful, simpler substance **C**.

By chance, clumps of organic molecules included the enzyme-like protein. Such clumps had an edge in the acquisition of starting materials. Suppose that the **C** molecules became scarce. The advantage tilted to

molecular clumps that promoted the formation of **C** from simpler substances **B** and **A**. Suppose that **B** and **A** were carbon dioxide and water. The atmosphere and seas contain unlimited amounts of both. Thus, chemical selection favored a synthetic pathway:

$$\textbf{A} + \textbf{B} \longrightarrow \textbf{C} \longrightarrow \textbf{D}$$

Were some clumps better at absorbing and using energy? Think back on chlorophyll *a* (Section 7.1). A group of rings in this pigment absorbs light and gives up electrons. The same kinds of ring structures occur in electron transfer chains in all photosynthetic and aerobically respiring cells. They form spontaneously from formaldehyde (Figure 20.5)—one of the legacies of cosmic clouds. Were similar structures transferring electrons in early metabolic pathways? Probably.

The point is, long before cells emerged, a form of chemical competition was under way. Enzymes and other reactive organic compounds had the competitive edge in the acquisition of energy and materials.

ORIGIN OF THE FIRST PLASMA MEMBRANES

All living cells have an outer membrane that controls which substances enter and leave the cytoplasm in a given interval (Section 5.1). By a current hypothesis, proto-cells were transitional forms between simple organic compounds and the first living cells. These **proto-cells** were no more than membrane-bound sacs

Figure 20.5 Hypothetical sequence of the chemical evolution of (**a**) an organic compound, formaldehyde, into (**c**) porphyrin.

Formaldehyde was present on the early Earth. Porphyrin is the light-absorbing and electron-donating part of chlorophyll molecules (**d**). It also is part of cytochrome, a protein component of the electron transfer chains in many metabolic pathways. It also is part of the heme of hemoglobin.

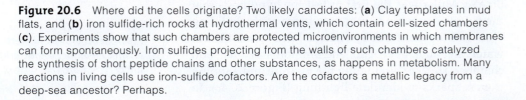

Figure 20.6 Where did the cells originate? Two likely candidates: (**a**) Clay templates in mud flats, and (**b**) iron sulfide-rich rocks at hydrothermal vents, which contain cell-sized chambers (**c**). Experiments show that such chambers are protected microenvironments in which membranes can form spontaneously. Iron sulfides projecting from the walls of such chambers catalyzed the synthesis of short peptide chains and other substances, as happens in metabolism. Many reactions in living cells use iron-sulfide cofactors. Are the cofactors a metallic legacy from a deep-sea ancestor? Perhaps.

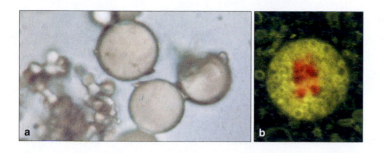

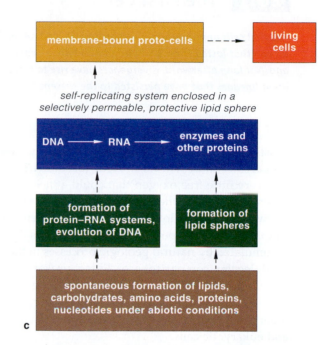

Figure 20.7 Laboratory-grown proto-cells. (**a**) Selectively permeable sacs. Heated amino acids formed protein chains. When moistened, the chains assembled into a membrane. (**b**) A membrane of fatty acids and alcohols (*green*) enclosing RNA-coated clay (*red*). The mineral-rich clay catalyzes RNA polymerization and promotes the formation of a membrane sac. (**c**) Model for steps in the chemical processes that led to the first living, self-replicating, membrane-bound cells.

that contained systems of enzymes and other agents of metabolism, and that were self-replicating.

Experiments reveal that membrane sacs can form spontaneously. Under conditions that simulate ancient sunbaked tidal flats, amino acids do form chains that surround a volume of fluid (Figure 20.7*a*). Fatty acids and alcohols spontaneously form vesicles, especially when clays rich in minerals are present (Figure 20.7*b*).

Or did proto-cells form from organic compounds at hydrothermal vents? Cell-sized chambers occur in mineral-rich rocks at existing vents (Figure 20.6*b,c*). Were the chamber walls replication templates for RNA, proteins, DNA, and lipids? The molecules would have accumulated inside, favoring the chemical conditions required for the emergence of living cells.

ORIGIN OF SELF-REPLICATING SYSTEMS

Life also is characterized by reproduction, which now starts with protein-building instructions in DNA. As you know from Section 14.3, it takes RNA, enzymes, and other molecules to translate DNA into proteins.

Coenzymes and metal ions assist most enzymes—and certain coenzymes are structurally identical with RNA subunits. When you mix and heat RNA subunits with very short chains of phosphate groups, they self-assemble into strands of RNA. Simple self-replicating systems of RNA, enzymes, and coenzymes have been made in laboratories. So we know RNA can serve as an information-storing template for making proteins.

Also, remember that one of the rRNA components of ribosomes catalyzes protein synthesis (Sections 14.3 and 14.4). The structure and function of ribosomes have been conserved over time; ribosomes of the most complex eukaryotes are extremely similar to those in prokaryotic cells of ancient lineages. rRNA's catalytic behavior probably evolved early in Earth history.

Did an **RNA world** *precede* the emergence of DNA? That is, were short RNA strands the first templates for protein synthesis? As you know, RNA and DNA are similar. Three of their four bases are identical. RNA's uracil differs from DNA's thymine by a single functional group. But DNA's *helically coiled, double-*

stranded structure is more stable than RNA, and it can store much more protein-building information in less space. There would have been selective advantage in functionally separating the storage of protein-building information (DNA) from protein synthesis (RNA).

Until we identify chemical ancestors of RNA and DNA, the history of life's origin will not be complete. But clues are coming in. For instance, researchers fed data about inorganic compounds and energy sources into a supercomputer. They programmed the computer to simulate random chemical reactions among organic compounds, which may well have happened untold billions of times in the distant past. Then they ran the program again and again.

The outcome of their experiment was always the same. *Simple precursors evolved. Then they spontaneously organized themselves into large, complex molecules. And they began to interact as complex systems.*

There are gaps in our knowledge of life's origin. But diverse laboratory experiments and computer simulations show that chemical processes can result in all organic molecules and structures that we think of as being characteristic of life.

20.3 The First Cells

LINKS TO
SECTIONS
4.3, 6.4, 7.8

The first cells apparently evolved during the Archaean, an eon that lasted from 3.8 billion to 2.5 billion years ago. Not long afterward, divergences gave rise to three great lineages that have persisted to the present.

THE GOLDEN AGE OF PROKARYOTES

Fossils indicate that the first cells were like existing prokaryotes; they had no nucleus (Section 4.3). There was very little free oxygen that could attack them. Its absence is a clue to their mode of nutrition. Anaerobic pathways would allow them to obtain energy from simple organic compounds and mineral ions that had accumulated by natural geologic processes in the seas.

Molecular comparisons of living prokaryotes tell us that some populations diverged not long after life originated. One lineage gave rise to the bacteria. The other gave rise to the shared ancestors of archaeans and eukaryotic cells.

Microscopically small fossils in 3.5-billion-year-old rocks give clues to what some of the first prokaryotes looked like (Figure 20.8a). Other fossils clearly show that chemoautotrophic forms had become established near deep-sea hydrothermal vents by 3.2 billion years ago. In some groups, pigments probably detected the type of weak infrared radiation (heat) that has been measured at hydrothermal vents. Pigments may have helped cells detect and avoid boiling water, as they do for some existing hydrothermal vent species.

Gene mutations arose independently in some of the prokaryotic populations. They led to modifications in radiation-sensitive pigments, electron transfer chains,

and other bits of metabolic machinery that started a novel mode of nutrition. We call it the cyclic pathway of photosynthesis. Those bacterial populations were photoautotrophic; they had tapped into sunlight, an unlimited energy source (Section 7.8).

As they reproduced, those self-feeding populations of tiny cells grew on top of one another. They became flattened mats, infiltrated with calcium carbonate and other dissolved mineral ions, and fine sediments. In time, they were transformed into dome-shaped fossils known as **stromatolites**. Radiometric dating tells us that some are 3 billion years old (Figure 20.9).

When the Proterozoic dawned 2.7 billion years ago, stromatolites were abundant. By that time, a noncyclic pathway of photosynthesis had evolved in a bacterial lineage, the cyanobacteria. Cyanobacterial populations increased, and so did the pathway's waste product—free oxygen. At first, oxygen slowly accumulated in the surface waters of the seas, then in air. So now we return to events sketched out in Chapter 7.

An atmosphere enriched with free oxygen had two irreversible effects. First, *it stopped the further chemical origin of living cells.* Except in a few anaerobic habitats, complex organic compounds could no longer assemble spontaneously and stay intact; they could not escape attacks by oxygen radicals. Second, *aerobic respiration evolved and in time became the dominant energy-releasing pathway.* In many prokaryotic lineages, selection had favored this pathway, which neutralized oxygen by *using* it as an electron acceptor. Aerobic respiration was a key innovation that contributed to the rise of all complex, multicelled eukaryotes.

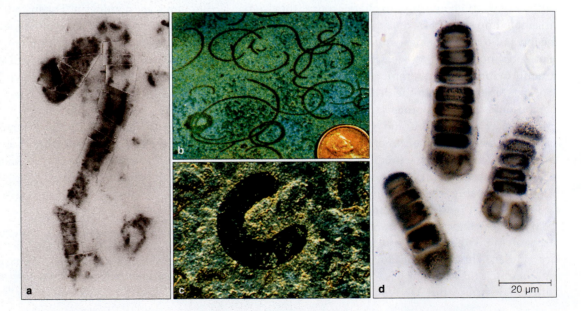

Figure 20.8 A sampling of early life. (**a**) A strand of what might be walled prokaryotic cells dates back 3.5 billion years. (**b**) One of the oldest known eukaryotic species, *Grypania spiralis*, which lived 2.1 billion years ago. Its fossilized colonies are large enough to see without a microscope. (**c**) Fossil of *Tawuia*, another early eukaryotic species that lived during the Proterozoic. (**d**) Fossils of a red alga, *Bangiomorpha pubescens*. This multicelled species lived 1.2 billion years ago, and it reproduced sexually.

20 μm

THE RISE OF EUKARYOTES

Eukaryotic cells also evolved during the Proterozoic. Traces of the kinds of lipids that existing eukaryotic cells produce have been isolated from rocks dated at 2.8 billion years old. But the first complete eukaryotic fossils are about 2.1 billion years old (Figure 20.8*b,c*). Those ancient species had organelles.

As you know, organelles are the defining features of eukaryotic cells. Where did they come from? The next section presents a few plausible hypotheses.

We still do not know how the earliest eukaryotes fit in evolutionary trees. The earliest known form we can assign to a modern group is the filamentous alga *Bangiomorpha pubescens*. This red alga, which lived 1.2 billion years ago, is the first multicelled eukaryotic species to be discovered. Its cells were differentiated. Some cells in its strandlike body served as anchoring structures. Others formed two types of sexual spores. Spore production certainly makes *B. pubescens* one of the earliest practitioners of sexual reproduction.

By 1.1 billion years ago the supercontinent Rodinia had formed. Stromatolites dotted its vast shorelines, but 300 million years later, they were in decline. Were the cyanobacteria a vast food source for predators and parasites? By then, protists, fungi, animals, and the algae that would later give rise to plants were sharing the shoreline with them. Also, 570 million years ago, when oxygen in the atmosphere approached modern levels, animals began their first adaptive radiations in the Cambrian seas. A coevolutionary arms race that continues to this day was off and running.

Figure 20.9 Some stromatolites. (**a**) A painting of how one shallow sea might have looked early in the Proterozoic.

(**b**) In Australia's Shark Bay are mounds that are 2,000 years old. They are structurally similar to stromatolites that formed 3 billion years ago.

(**c**) A cut stromatolite reveals many layers of fine sediments and mineral deposits. The cyanobacterial cells often were preserved as well.

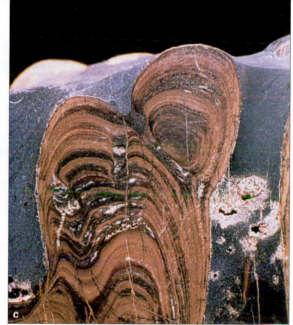

The first living cells evolved by 3.8 billion years ago, in the Archaean eon. All were prokaryotic, and they obtained energy by anaerobic pathways. Not long afterward, the ancestors of archaeans and eukaryotic cells diverged from the lineage that led to modern bacteria.

After the noncyclic pathway of photosynthesis evolved, free oxygen accumulated in the atmosphere and ended the further spontaneous chemical origin of life. The stage was set for the evolution of eukaryotic cells.

20.4 Where Did Organelles Come From?

LINKS TO
SECTIONS
4.3, 4.5–4.8, 14.2

Thanks to globe-hopping microfossil hunters, we have considerable evidence of early life, including the fossil treasures shown in Sections 4.3 and 20.3. Today, most descendant species contain a profusion of organelles. Where did the organelles come from?

ORIGIN OF THE NUCLEUS AND ER

Prokaryotic cells, recall, do not have an abundance of organelles. Some do have infoldings of their plasma membrane, which incorporates many enzymes and other components used in metabolic reactions (Figure 20.10a). Applying the theory of natural selection, we may hypothesize that infoldings originated among ancestors of eukaryotic cells. What advantages did the infoldings offer? They became channels that could concentrate nutrients, organic compounds, and other substances. Also, a membrane with a greater surface area could be a physical platform for more metabolic

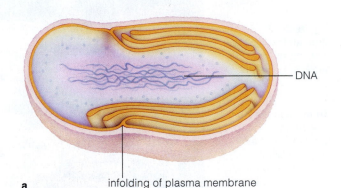

a

infolding of plasma membrane

DNA

b

Figure 20.10 (**a**) Sketch of a bacterial cell (*Nitrobacter*) that lives in soil. Cytoplasmic fluid bathes permanent infoldings of the plasma membrane. (**b**) Model for the origin of the nuclear envelope and the endoplasmic reticulum. In prokaryotic ancestors of eukaryotic cells, infoldings of the plasma membrane may have evolved into these organelles.

machinery as well as transport proteins. Remember the surface-to-volume ratio?

The channels of endoplasmic reticulum (ER) may have evolved this way. They also may have protected the metabolic machinery from uninvited guests. From time to time, metabolically "hungry" foreign cells do enter the cytoplasm of existing prokaryotic cells.

Some infoldings might have extended around the DNA, the start of a nuclear envelope (Figure 20.10b). A nuclear envelope would have been favored because it helped protect the cell's hereditary material from foreign DNA. Bacteria and the simple eukaryotic cells called yeasts can transfer plasmids among themselves. Early eukaryotic cells with a nuclear envelope could copy and use their messages of inheritance, free from metabolic competition from a potentially disruptive hodgepodge of foreign DNA.

ORIGIN OF MITOCHONDRIA AND CHLOROPLASTS

Early in the history of life, cells became food for one another. Heterotrophs engulfed autotrophs and other heterotrophs. Intracellular parasites dined inside their hosts. In some cases, the engulfed meals or parasites struck an uneasy balance with the host cells. They were protected, they withdrew some nutrients from the cytoplasm, and—like their hosts—they continued to divide and reproduce. Over time, they evolved into mitochondria, chloroplasts, and some other organelles.

The novel partnerships are one premise of a theory of **endosymbiosis**, as championed by Lynn Margulis and others. (*Endo–* means within and *symbiosis* means living together.) The symbiont species lives out its life inside a host species, and the interaction benefits one or both of them.

By this theory, eukaryotic cells evolved after the noncyclic pathway of photosynthesis emerged and permanently changed the atmosphere. By 2.1 billion years ago, remember, certain prokaryotic cells had adapted to the concentration of free oxygen and were already engaged in aerobic respiration. The ancestors of eukaryotic cells preyed upon some aerobic bacteria and were parasitized by others (Figure 20.11a). At that time, endosymbiotic interactions began.

The host began to use ATP produced by its aerobic symbiont. The aerobe no longer had to spend energy on acquiring raw materials; the host did this work for it. DNA regions that specified proteins produced by both host and symbiont were free to mutate and lose their function in one partner or the other. In time, both types of cells became incapable of independent life.

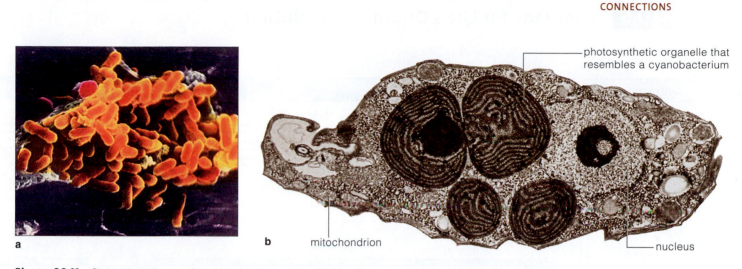

Figure 20.11 Clues to ancient endosymbiotic interactions. (**a**) What the ancestors of mitochondria may have looked like. The protist *Reclinomonas americana* has the structurally simplest mitochondria. The mitochondrial genes resemble genes of *Rickettsia prowazekii*, a parasitic bacterium that causes typhus. Like mitochondria, *R. prowazekii* divides only inside the cytoplasm of eukaryotic cells. Enzymes in the cytoplasm catalyze the partial breakdown of organic compounds—a task that is completed inside aerobically respiring mitochondria. (**b**) *Cyanophora paradoxa* is one of the flagellated protists called glaucophytes. Its mitochondria resemble aerobic bacteria in size and structure. Its photosynthetic structures resemble cyanobacteria—they even have a wall like that of cyanobacteria.

EVIDENCE OF ENDOSYMBIOSIS

Is such a theory far-fetched? A chance discovery in Jeon Kwang's laboratory suggests otherwise. In 1966, a rod-shaped bacterium had infected his culture of *Amoeba discoides*. Some infected cells died right away. Others grew more slowly, and they were smaller and vulnerable to starving to death. Kwang maintained the infected culture. Five years later, infected amoebas were harboring many bacterial cells, yet they were all thriving. Exposure to antibiotics killed the bacterial cells (but not the amoebas).

Infection-free cells were stripped of their nucleus and got a nucleus from an infected cell. They died. Yet more than 90 percent survived when a few bacteria were included with the transplant. As other studies showed, the infected amoebas had lost their ability to synthesize an essential enzyme. They depended on the bacterium to make it for them! Invading bacterial cells had become symbiotic with the amoebas.

When you think about it, mitochondria do resemble bacteria in size and structure. Each has its own DNA and divides independently of cell division. The inner membrane of a mitochondrion resembles a bacterial cell's plasma membrane. Its DNA has just a few genes (thirty-seven in human mitochondrial DNA). Also, a few of the codons are slightly different from those of the near-universal genetic code (Section 14.2).

We can predict that chloroplasts, too, originated by endosymbiosis. In one scenario, photosynthetic cells were engulfed by predatory aerobic bacteria, but they escaped digestion. They started to absorb nutrients in the host's cytoplasm and continued to function. They also released oxygen when they photosynthesized. By releasing oxygen inside the aerobically respiring hosts, they acted as agents favoring endosymbiosis.

In their metabolism and their overall nucleic acid sequence, existing chloroplasts resemble cyanobacteria. The chloroplast DNA replicates itself independently of cellular DNA. Chloroplasts and the cells in which they reside divide independently of each other.

Or consider the protists called glaucophytes. They have unique photosynthetic organelles that resemble cyanobacteria. These organelles even have their own cell wall (Figure 20.11*b*).

However they arose, the first eukaryotic cells had a nucleus, an endomembrane system, mitochondria and, in some lineages, chloroplasts. They were the world's first protists. They had efficient metabolic systems, and they evolved fast. In no time at all, evolutionarily speaking, some of their descendants evolved into the plants, fungi, and animals. The next section provides a time frame for these pivotal events.

A nucleus and other organelles are defining features of eukaryotic cells. The nucleus and ER may have evolved by infoldings of the plasma membrane. Mitochondria and chloroplasts may have evolved through endosymbiosis between heterotrophic host cells and their prey or parasites.

20.5 Time Line for Life's Origin and Evolution

LINKS TO
SECTIONS
1.1, 17.5

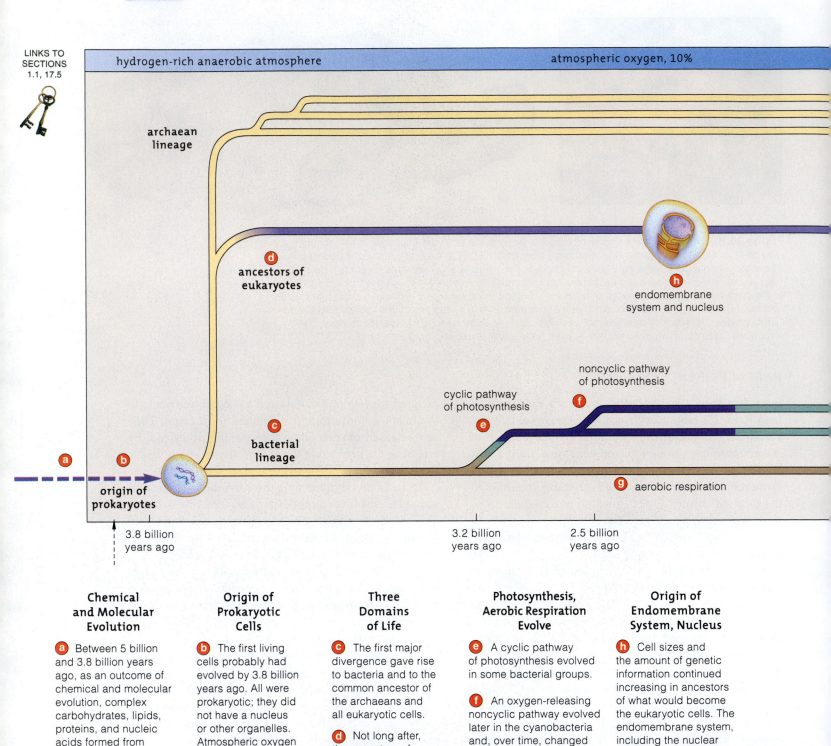

Chemical and Molecular Evolution

(a) Between 5 billion and 3.8 billion years ago, as an outcome of chemical and molecular evolution, complex carbohydrates, lipids, proteins, and nucleic acids formed from the simple organic compounds present on the early Earth.

Origin of Prokaryotic Cells

(b) The first living cells probably had evolved by 3.8 billion years ago. All were prokaryotic; they did not have a nucleus or other organelles. Atmospheric oxygen was low and the early cells made ATP by anaerobic pathways.

Three Domains of Life

(c) The first major divergence gave rise to bacteria and to the common ancestor of the archaeans and all eukaryotic cells.

(d) Not long after, the ancestors of archaeans and eukaryotic cells diverged.

Photosynthesis, Aerobic Respiration Evolve

(e) A cyclic pathway of photosynthesis evolved in some bacterial groups.

(f) An oxygen-releasing noncyclic pathway evolved later in the cyanobacteria and, over time, changed the atmosphere.

(g) Aerobic respiration evolved independently in many bacterial groups.

Origin of Endomembrane System, Nucleus

(h) Cell sizes and the amount of genetic information continued increasing in ancestors of what would become the eukaryotic cells. The endomembrane system, including the nuclear envelope, arose through the modification of cell membranes.

Figure 20.12 *Animated!* Milestones in the history of life. As you read the next unit on life's past and present diversity, refer to this visual overview. It can serve as a simple reminder of the evolutionary connections among all groups of organisms, from the structurally simple to the most complex.

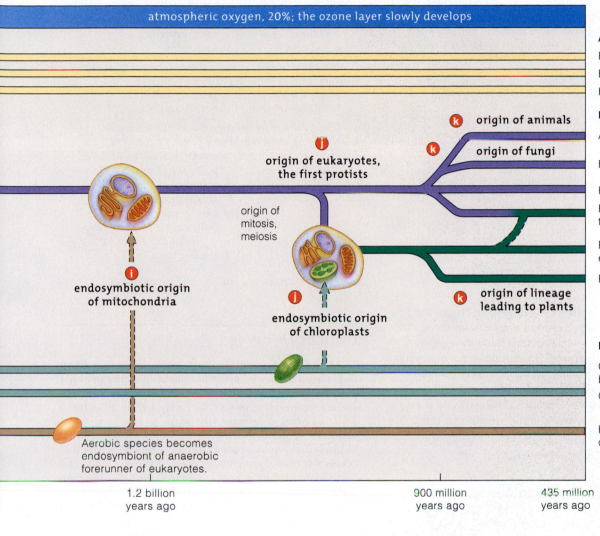

atmospheric oxygen, 20%; the ozone layer slowly develops

ARCHAEA

Extreme thermophiles

Extreme thermophiles and mesophiles

Halophiles and methanogens

EUKARYA

Animals

Fungi

Heterotrophic protists

Photosynthetic protists with chloroplasts that evolved from red and green algae

Red and green algae; their chloroplasts evolved from cyanobacterial symbionts

Plants

BACTERIA

Oxygen-releasing photosynthetic bacteria (cyanobacteria)

Other photosynthetic bacteria

Heterotrophic bacteria, including chemoheterotrophs

k origin of animals

j origin of eukaryotes, the first protists

k origin of fungi

origin of mitosis, meiosis

i endosymbiotic origin of mitochondria

j endosymbiotic origin of chloroplasts

k origin of lineage leading to plants

Aerobic species becomes endosymbiont of anaerobic forerunner of eukaryotes.

1.2 billion years ago   900 million years ago   435 million years ago

Endosymbiotic Origin of Mitochondria

i Before about 1.2 billion years ago, aerobic bacterial species and an anaerobic ancestor of eukaryotic cells entered into close symbiotic interaction. The endosymbiont evolved into the mitochondrion.

Endosymbiotic Origin of Chloroplasts

j Cyanobacteria entered into a close symbiotic interaction with early protists and evolved into chloroplasts. Later, photosynthetic protists would evolve into chloroplasts inside other protist hosts.

Plants, Fungi, and Animals Evolve

k By 900 million years ago, all major lineages—including fungi, animals, and the algae that would give rise to plants—had evolved along shorelines of the first supercontinent.

Lineages That Have Endured to the Present

l Today, organisms live in all regions of Earth's waters, crust, and atmosphere. They are related by descent and share certain traits. However, each lineage encountered different selective pressures, and each has evolved its own characteristic traits.

Summary

Section 20.1 Earth formed more than 4 billion years ago. Experimental tests, information on the formation of stars and planets, and other lines of research offer indirect evidence that the complex organic compounds characteristic of life could have formed spontaneously under the conditions that prevailed on the early Earth.

Biology Now
See experiments on how organic compounds can form spontaneously with the animation on BiologyNow.

Section 20.2 The emergence of the first cells was preceded by chemical evolution that led to enzymes and other agents of metabolism, the self-assembly of membranes on environmental templates, and a self-replicating system. RNA probably was the template for protein synthesis before DNA evolved as an efficient way to store protein-building information.

Biology Now
Read the InfoTrac article "Transitions from Nonliving to Living Matter," Steen Rasmussen et al., Science, February 2004. Also "First Cell," David Deamer, Discover, November 1995.

Section 20.3 The first cells may have originated 3.8 billion years ago. They were anaerobic prokaryotes. An early divergence separated bacteria from the ancestors of archaeans and eukaryotes. Evolution of the noncyclic pathway of photosynthesis in cyanobacteria resulted in an accumulation of free oxygen in the atmosphere, which favored aerobic respiration. This pathway was a key innovation in the evolution of eukaryotic cells.

Biology Now
Explore levels of biological organization with the interaction on BiologyNow.

Section 20.4 The internal membranes of eukaryotic cells may have evolved through infoldings of the cell membrane. Mitochondria and chloroplasts most likely evolved by endosymbiosis, at the times indicated in the Section 20.5 visual summary.

Section 20.5 Key events in life's origin and early evolution can be correlated with the geologic time scale.

Biology Now
Investigate the history of life with the animated interaction on BiologyNow.

Self-Quiz
Answers in Appendix II

1. An abundance of _____ in the atmosphere would have prevented the spontaneous (abiotic) assembly of organic compounds on the early Earth.
   a. hydrogen   b. methane   c. oxygen   d. nitrogen

2. The prevalence of iron-sulfide cofactors in living organisms may be evidence that life arose _____ .
   a. in outer space   c. near deep-sea vents
   b. on tidal flats   d. in the upper atmosphere

3. The evolution of _____ resulted in an increase in the levels of atmospheric oxygen.
   a. sexual reproduction
   b. aerobic respiration
   c. the noncyclic pathway of photosynthesis
   d. the cyclic pathway of photosynthesis

4. Mitochondria may have evolved from _____ .
   a. chloroplasts   c. early protists
   b. bacteria   d. archaeans

5. Infoldings of the plasma membrane into the cytoplasm of some prokaryotes may have evolved into the _____ .
   a. nuclear envelope   c. primary cell wall
   b. ER membranes   d. both a and b

6. Chronologically arrange the evolutionary events, with 1 being the earliest and 6 the most recent.
   ____ 1   a. emergence of the noncyclic
   ____ 2   pathway of photosynthesis
   ____ 3   b. origin of mitochondria
   ____ 4   c. origin of proto-cells
   ____ 5   d. emergence of the cyclic
   ____ 6   pathway of photosynthesis
             e. origin of chloroplasts
             f. the big bang

Additional questions are available on **Biology Now™**

Critical Thinking

1. Mars formed about 5 million years earlier than Earth, and it has a similar composition but is far richer in iron. It is farther from the sun and much chillier, with an average surface temperature of −63° C. Today, nearly all of the water on Mars is permanently frozen in soil. To some researchers, photographs of certain geological features indicate that liquid water might have flowed across the planet's surface during an earlier and warmer time. The Martian atmosphere is now richer in carbon dioxide than Earth's, but very low in nitrogen and oxygen. Based on this information, would you rule out the possibility that life could have existed on Mars or that simple life forms could currently exist there? Explain your reasoning.

2. What if it were possible to create life in test tubes? That is the idea behind modeling and perhaps creating minimal organisms: living cells having the smallest set of genes required to survive and reproduce.

   Craig Venter and Claire Fraser found that *Mycoplasma genitalium*, a bacterium that has only 517 genes (and 2,209 transposons), is a good candidate for such an experiment. By disabling its genes one at a time, they discovered it may have only 265–350 essential protein-coding genes.

   What if those genes were synthesized one at a time and inserted into an engineered cell consisting only of a plasma membrane and cytoplasm? Would the cell come to life? The possibility that it might prompted Venter and Fraser to seek advice from a panel of bioethicists and theologians. No one on the panel objected to synthetic life research. They said that much good might come of it, provided scientists did not claim to have found "the secret of life." The December 10, 1999, issue of *Science* includes an essay from the panel and an article on *M. genitalium* research. Read both, then write down your thoughts about "creating" life in a test tube.

Appendix I. Classification System

This revised classification scheme is a composite of several that microbiologists, botanists, and zoologists use. The major groupings are agreed upon, more or less. However, there is not always agreement on what to name a particular grouping or where it might fit within the overall hierarchy. There are several reasons why full consensus is not possible at this time.

First, the fossil record varies in its completeness and quality. Therefore, the phylogenetic relationship of one group to other groups is sometimes open to interpretation. Today, comparative studies at the molecular level are firming up the picture, but the work is still under way. Also, molecular comparisons do not always provide definitive answers to questions about phylogeny. Comparisons based on one set of genes may conflict with those comparing a different part of the genome. Or comparisons with one member of a group may conflict with comparisons based on other group members.

Second, ever since the time of Linnaeus, systems of classification have been based on the perceived morphological similarities and differences among organisms. Although some original interpretations are now open to question, we are so used to thinking about organisms in certain ways that reclassification often proceeds slowly.

A few examples: Traditionally, birds and reptiles were grouped in separate classes (Reptilia and Aves); yet there are compelling arguments for grouping the lizards and snakes in one group and the crocodilians, dinosaurs, and birds in another. Many biologists still favor a six-kingdom system of classification (archaea, bacteria, protists, plants, fungi, and animals). Others advocate a switch to the more recently proposed three-domain system (archaea, bacteria, and eukarya).

Third, researchers in microbiology, mycology, botany, zoology, and other fields of inquiry inherited a wealth of literature, based on classification systems that have been developed over time in each field of inquiry. Many are reluctant to give up established terminology that offers access to the past.

For example, botanists and microbiologists often use *division*, and zoologists *phylum*, for taxa that are equivalent in hierarchies of classification.

Why bother with classification frameworks if we know they only imperfectly reflect the evolutionary history of life? We do so for the same reasons that a writer might break up a history of civilization into several volumes, each with a number of chapters. Both are efforts to impart structure to an enormous body of knowledge and to facilitate retrieval of information from it. More importantly, to the extent that modern classification schemes accurately reflect evolutionary relationships, they provide the basis for comparative biological studies, which link all fields of biology.

Bear in mind that we include this appendix for your reference purposes only. Besides being open to revision, it is not meant to be complete. Names shown in "quotes" are polyphyletic or paraphyletic groups that are undergoing revision. For example, "reptiles" comprise at least three and possibly more lineages.

The most recently discovered species, as from the mid-ocean province, are not listed. Many existing and extinct species of the more obscure phyla are also not represented. Our strategy is to focus primarily on the organisms mentioned in the text or familiar to most students. We delve more deeply into flowering plants than into bryophytes, and into chordates than annelids.

PROKARYOTES AND EUKARYOTES COMPARED

As a general frame of reference, note that almost all bacteria and archaea are microscopic in size. Their DNA is concentrated in a nucleoid (a region of cytoplasm), not in a membrane-bound nucleus. All are single cells or simple associations of cells. They reproduce by prokaryotic fission or budding; they transfer genes by bacterial conjugation.

Table A lists representative types of autotrophic and heterotrophic prokaryotes. The authoritative reference, *Bergey's Manual of Systematic Bacteriology*, has called this a time of taxonomic transition. It references groups mainly by numerical taxonomy (Section 19.1) rather than by phylogeny. Our classification system does reflect evidence of evolutionary relationships for at least some bacterial groups.

The first life forms were prokaryotic. Similarities between Bacteria and Archaea have more ancient origins relative to the traits of eukaryotes.

Unlike the prokaryotes, all eukaryotic cells start out life with a DNA-enclosing nucleus and other membrane-bound organelles. Their chromosomes have many histones and other proteins attached. They include spectacularly diverse single-celled and multicelled species, which can reproduce by way of meiosis, mitosis, or both.

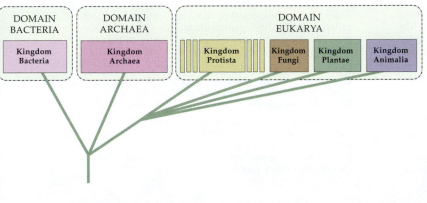

DOMAIN OF BACTERIA

KINGDOM BACTERIA   The largest, and most diverse group of prokaryotic cells. Includes photosynthetic autotrophs, chemosynthetic autotrophs, and heterotrophs. All prokaryotic pathogens of vertebrates are bacteria.

PHYLUM AQIFACAE   Most ancient branch of the bacterial tree. Gram-negative, mostly aerobic chemoautotrophs, mainly of volcanic hot springs. *Aquifex.*

PHYLUM DEINOCOCCUS-THERMUS   Gram-positive, heat-loving chemoautotrophs. *Deinococcus* is the most radiation resistant organism known. *Thermus* occurs in hot springs and near hydrothermal vents.

PHYLUM CHLOROFLEXI   Green nonsulfur bacteria. Gram-negative bacteria of hot springs, freshwater lakes, and marine habitats. Act as nonoxygen-producing photoautotrophs or aerobic chemoheterotrophs. *Chloroflexus.*

PHYLUM ACTINOBACTERIA   Gram-positive, mostly aerobic heterotrophs in soil, freshwater and marine habitats, and on mammalian skin. *Propionibacterium, Actinomyces, Streptomyces.*

PHYLUM CYANOBACTERIA   Gram-negative, oxygen-releasing photoautotrophs mainly in aquatic habitats. They have chlorophyll *a* and photosystem I. Includes many nitrogen-fixing genera. *Anabaena, Nostoc, Oscillatoria.*

PHYLUM CHLOROBIUM   Green sulfur bacteria. Gram-negative nonoxygen-producing photosynthesizers, mainly in freshwater sediments. *Chlorobium.*

PHYLUM FIRMICUTES   Gram-positive walled cells and the cell wall-less mycoplasmas. All are heterotrophs. Some survive in soil, hot springs, lakes, or oceans. Others live on or in animals. *Bacillus, Clostridium, Heliobacterium, Lactobacillus, Listeria, Mycobacterium, Mycoplasma, Streptococcus.*

PHYLUM CHLAMYDIAE   Gram-negative intracellular parasites of birds and mammals. *Chlamydia.*

PHYLUM SPIROCHETES   Free-living, parasitic, and mutualistic gram-negative spring-shaped bacteria. *Borelia, Pillotina, Spirillum, Treponema.*

PHYLUM PROTEOBACTERIA   The largest bacterial group. Includes photoautotrophs, chemoautotrophs, and heterotrophs; free-living, parasitic, and colonial groups. All are gram-negative.

Class Alphaproteobacteria. *Agrobacterium, Azospirillum, Nitrobacter, Rickettsia, Rhizobium.*

Class Betaproteobacteria. *Neisseria.*

Class Gammaproteobacteria. *Chromatium, Escherichia, Haemopilius, Pseudomonas, Salmonella, Shigella, Thiomargarita, Vibrio, Yersinia.*

Class Deltaproteobacteria. *Azotobacter, Myxococcus.*

Class Epsilonproteobacteria. *Campylobacter, Helicobacter.*

DOMAIN OF ARCHAEA

KINGDOM ARCHAEA   Prokaryotes that are evolutionarily between eukaryotic cells and the bacteria. Most are anaerobes. None are photosynthetic. Originally discovered in extreme habitats, they are now known to be widely dispersed. Compared with bacteria, the archaea have a distinctive cell wall structure and unique membrane lipids, ribosomes, and RNA sequences. Some are symbiotic with animals, but none are known to be animal pathogens.

PHYLUM EURYARCHAEOTA   Largest archean group. Includes extreme thermophiles, halophiles, and methanogens. Others are abundant in the upper waters of the ocean and other more moderate habitats. *Methanocaldococcus, Nanoarchaeum.*

PHYLUM CRENARCHAEOTA   Includes extreme theromophiles, as well as species that survive in Antarctic waters, and in more moderate habitats. *Sulfolobus, Ignicoccus.*

PHYLUM KORARCHAEOTA   Known only from DNA isolated from hydrothermal pools. As of this writing, none have been cultured and no species have been named.

DOMAIN OF EUKARYOTES

KINGDOM "PROTISTA"   A collection of single-celled and multicelled lineages, which does not constitute a monophyletic group. Some biologists consider the groups listed below to be kingdoms in their own right.

PARABASALIA   Parabasalids. Flagellated, single-celled anaerobic heterotrophs with a cytoskeletal "backbone" that runs the length of the cell. There are no mitochondria, but a hydrogenosome serves a similar function. *Trichomonas, Trichonympha.*

DIPLOMONADIDA   Diplomonads. Flagellated, anaerobic single-celled heterotrophs that do not have mitochondria or Golgi bodies and do not form a bipolar spindle at mitosis. May be one of the most ancient lineages. *Giardia.*

EUGLENOZOA   Euglenoids and kinetoplastids. Free-living and parasitic flagellates. All with one or more mitochondria. Some photosynthetic euglenoids with chloroplasts, others heterotrophic. *Euglena, Trypanosoma, Leishmania.*

RHIZARIA   Formaminiferans and radiolarians. Free-living, heterotrophic amoeboid cells that are enclosed in shells. Most live in ocean waters or sediments. *Pterocorys, Stylosphaera.*

ALVEOLATA   Single cells having a unique array of membrane-bound sacs (alveoli) just beneath the plasma membrane.

Ciliata. Ciliated protozoans. Heterotrophic protists with many cilia. *Paramecium, Didinium.*

Dinoflagellates. Diverse heterotrophic and photosynthetic flagellated cells that deposit cellulose in their alveoli. *Gonyaulax, Gymnodinium, Karenia, Noctiluca.*

Apicomplexans. Single-celled parasites of animals. A unique microtubular device is used to attach to and penetrate a host cell. *Plasmodium.*

STRAMENOPHILA   Stramenophiles. Single-celled and multicelled forms; flagella with tinsel-like filaments.

Oomycotes. Water molds. Heterotrophs. Decomposers, some parasites. *Saprolegnia, Phytophthora, Plasmopara.*

Chrysophytes. Golden algae, yellow-green algae, diatoms, coccolithophores. Photosynthetic. *Emiliania, Mischococcus.*

Phaeophytes. Brown algae. Photosynthetic; nearly all live in temperate marine waters. All are multicellular. *Macrocystis, Laminaria, Sargassum, Postelsia.*

RHODOPHYTA   Red algae. Mostly photosynthetic, some parasitic. Nearly all marine, some in freshwater habitats. Most multicellular. *Porphyra, Antithamion.*

CHLOROPHYTA   Green algae. Mostly photosynthetic, some parasitic. Most freshwater, some marine or terrestrial. Single-celled, colonial, and multicellular forms. Some biologists place the chlorophytes and charophytes with the land plants in a kingdom called the Viridiplantae. *Acetabularia, Chlamydomonas, Chlorella, Codium, Udotea, Ulva, Volvox.*

CHAROPHYTA   Photosynthetic. Closest living relatives of plants. Include both single-celled and multicelled forms. Desmids, stoneworts. *Micrasterias, Chara, Spirogyra.*

AMOEBOZOA   True amoebas and slime molds. Heterotrophs that spend all or part of the life cycle as a single cell that uses pseudopods to capture food. *Amoeba, Entoamoeba* (amoebas), *Dictyostelium* (cellular slime mold), *Physarum* (plasmodial slime mold).

KINGDOM FUNGI

Nearly all multicelled eukaryotic species with chitin-containing cell walls. Heterotrophs, mostly saprobic decomposers, some parasites. Nutrition based upon extracellular digestion of organic matter and absorption of nutrients by individual cells. Multicelled species form absorptive mycelia and reproductive structures that produce asexual spores (and sometimes sexual spores).

PHYLUM CHYTRIDIOMYCOTA Chytrids. Primarily aquatic; saprobic decomposers or parasites that produce flagellated spores. *Chytridium.*

PHYLUM ZYGOMYCOTA Zygomycetes. Producers of zygospores (zygotes inside thick wall) by way of sexual reproduction. Bread molds, related forms. *Rhizopus, Philobolus.*

PHYLUM ASCOMYCOTA Ascomycetes. Sac fungi. Sac-shaped cells form sexual spores (ascospores). Most yeasts and molds, morels, truffles. *Saccharomycetes, Morchella, Neurospora, Claviceps, Candida, Aspergillus, Penicillium.*

PHYLUM BASIDIOMYCOTA Basidiomycetes. Club fungi. Most diverse group. Produce basidiospores inside club-shaped structures. Mushrooms, shelf fungi, stinkhorns. *Agaricus, Amanita, Craterellus, Gymnophilus, Puccinia, Ustilago.*

"IMPERFECT FUNGI" Sexual spores absent or undetected. The group has no formal taxonomic status. If better understood, a given species might be grouped with sac fungi or club fungi. *Arthobotrys, Histoplasma, Microsporum, Verticillium.*

"LICHENS" Mutualistic interactions between fungal species and a cyanobacterium, green alga, or both. *Lobaria, Usnea.*

KINGDOM PLANTAE

Most photosynthetic with chlorophylls *a* and *b*. Some parasitic. Nearly all live on land. Sexual reproduction predominates.

BRYOPHYTES (NONVASCULAR PLANTS)

Small flattened haploid gametophyte dominates the life cycle; sporophyte remains attached to it. Sperm are flagellated; require water to swim to eggs for fertilization.

PHYLUM HEPATOPHYTA Liverworts. *Marchantia.*

PHYLUM ANTHOCEROPHYTA Hornworts.

PHYLUM BRYOPHYTA Mosses. *Polytrichum, Sphagnum.*

SEEDLESS VASCULAR PLANTS

Diploid sporophyte dominates, free-living gametophytes, flagellated sperm require water for fertilization.

PHYLUM LYCOPHYTA Lycophytes, club mosses. Small single-veined leaves, branching rhizomes. *Lycopodium, Selaginella.*

PHYLUM MONILOPHYTA

Subphylum Psilophyta. Whisk ferns. No obvious roots or leaves on sporophyte, very reduced. *Psilotum.*

Subphylum Sphenophyta. Horsetails. Reduced scalelike leaves. Some stems photosynthetic, others spore-producing. *Calamites* (extinct), *Equisetum.*

Subphylum Pterophyta. Ferns. Large leaves, usually with sori. Largest group of seedless vascular plants (12,000 species), mainly tropical, temperate habitats. *Pteris, Trichomanes, Cyathea* (tree ferns), *Polystichum.*

SEED-BEARING VASCULAR PLANTS

PHYLUM CYCADOPHYTA Cycads. Group of gymnosperms (vascular, bear "naked" seeds). Tropical, subtropical. Compound leaves, simple cones on male and female plants. Plants usually palm-like. Motile sperm. *Zamia, Cycas.*

PHYLUM GINKGOPHYTA Ginkgo (maidenhair tree). Type of gymnosperm. Motile sperm. Seeds with fleshy layer. *Ginkgo.*

PHYLUM GNETOPHYTA Gnetophytes. Only gymnosperms with vessels in xylem and double fertilization (but endosperm does not form). *Ephedra, Welwitchia, Gnetum.*

PHYLUM CONIFEROPHYTA Conifers. Most common and familiar gymnosperms. Generally cone-bearing species with needle-like or scale-like leaves. Includes pines (*Pinus*), redwoods (*Sequoia*), yews (*Taxus*).

PHYLUM ANTHOPHYTA Angiosperms (the flowering plants). Largest, most diverse group of vascular seed-bearing plants. Only organisms that produce flowers, fruits. Some families from several representative orders are listed:

BASAL FAMILIES

Family Amborellaceae. *Amborella.*
Family Nymphaeaceae. Water lilies.
Family Illiciaceae. Star anise.

MAGNOLIIDS

Family Magnoliaceae. Magnolias.
Family Lauraceae. Cinnamon, sassafras, avocados.
Family Piperaceae. Black pepper, white pepper.

EUDICOTS

Family Papaveraceae. Poppies.
Family Cactaceae. Cacti.
Family Euphorbiaceae. Spurges, poinsettia.
Family Salicaceae. Willows, poplars.
Family Fabaceae. Peas, beans, lupines, mesquite.
Family Rosaceae. Roses, apples, almonds, strawberries.
Family Moraceae. Figs, mulberries.
Family Cucurbitaceae. Squashes, melons, cucumbers.
Family Fagaceae. Oaks, chestnuts, beeches.
Family Brassicaceae. Mustards, cabbages, radishes.
Family Malvaceae. Mallows, okra, cotton, hibiscus, cocoa.
Family Sapindaceae. Soapberry, litchi, maples.
Family Ericaceae. Heaths, blueberries, azaleas.
Family Rubiaceae. Coffee.
Family Lamiaceae. Mints.
Family Solanaceae. Potatoes, eggplant, petunias.
Family Apiaceae. Parsleys, carrots, poison hemlock.
Family Asteraceae. Composites. Chrysanthemums, sunflowers, lettuces, dandelions.

MONOCOTS

Family Araceae. Anthuriums, calla lily, philodendrons.
Family Liliaceae. Lilies, tulips.
Family Alliaceae. Onions, garlic.
Family Iridaceae. Irises, gladioli, crocuses.
Family Orchidaceae. Orchids.
Family Arecaceae. Date palms, coconut palms.
Family Bromeliaceae. Bromeliads, pineapples.
Family Cyperaceae. Sedges.
Family Poaceae. Grasses, bamboos, corn, wheat, sugarcane.
Family Zingiberaceae. Gingers.

KINGDOM ANIMALIA

Multicelled heterotrophs, nearly all with tissues and organs, and organ systems, that are motile during part of the life cycle. Sexual reproduction occurs in most, but some also reproduce asexually. Embryos develop through a series of stages.

PHYLUM PORIFERA Sponges. No symmetry, tissues.

PHYLUM PLACOZOA Marine. Simplest known animal. Two cell layers, no mouth, no organs. *Trichoplax.*

PHYLUM CNIDARIA Radial symmetry, tissues, nematocysts.
Class Hydrozoa. Hydrozoans. *Hydra, Obelia, Physalia, Prya.*
Class Scyphozoa. Jellyfishes. *Aurelia.*
Class Anthozoa. Sea anemones, corals. *Telesto.*

PHYLUM PLATYHELMINTHES   Flatworms. Bilateral, cephalized; simplest animals with organ systems. Saclike gut.

Class Turbellaria. Triclads (planarians), polyclads. *Dugesia*.
Class Trematoda. Flukes. *Clonorchis, Schistosoma*.
Class Cestoda. Tapeworms. *Diphyllobothrium, Taenia*.

PHYLUM ROTIFERA   Rotifers. *Asplancha, Philodina*.

PHYLUM MOLLUSCA   Mollusks.

Class Polyplacophora. Chitons. *Cryptochiton, Tonicella*.

Class Gastropoda. Snails, sea slugs, land slugs. *Aplysia, Ariolimax, Cypraea, Haliotis, Helix, Liguus, Limax, Littorina*.

Class Bivalvia. Clams, mussels, scallops, cockles, oysters, shipworms. *Ensis, Chlamys, Mytelus, Patinopectin*.

Class Cephalopoda. Squids, octopuses, cuttlefish, nautiluses. *Dosidiscus, Loligo, Nautilus, Octopus, Sepia*.

PHYLUM ANNELIDA   Segmented worms.

Class Polychaeta. Mostly marine worms. *Eunice, Neanthes*.

Class Oligochaeta. Mostly freshwater and terrestrial worms, many marine. *Lumbricus* (earthworms), *Tubifex*.

Class Hirudinea. Leeches. *Hirudo, Placobdella*.

PHYLUM NEMATODA   Roundworms. *Ascaris, Caenorhabditis elegans, Necator* (hookworms), *Trichinella*.

PHYLUM ARTHROPODA

Subphylum Chelicerata. Chelicerates. Horseshoe crabs, spiders, scorpions, ticks, mites.

Subphylum Crustacea. Shrimps, crayfishes, lobsters, crabs, barnacles, copepods, isopods (sowbugs).

Subphylum Myriapoda. Centipedes, millipedes.

Subphylum Hexapoda. Insects and sprintails.

PHYLUM ECHINODERMATA   Echinoderms.

Class Asteroidea. Sea stars. *Asterias*.
Class Ophiuroidea. Brittle stars.
Class Echinoidea. Sea urchins, heart urchins, sand dollars.
Class Holothuroidea. Sea cucumbers.
Class Crinoidea. Feather stars, sea lilies.
Class Concentricycloidea. Sea daisies.

PHYLUM CHORDATA   Chordates.

Subphylum Urochordata. Tunicates, related forms.
Subphylum Cephalochordata. Lancelets.

CRANIATES

Class Myxini. Hagfishes.

VERTEBRATES (SUBGROUP OF CRANIATES)

Class Cephalaspidomorphi. Lampreys.

Class Chondrichthyes. Cartilaginous fishes (sharks, rays, skates, chimaeras).

Class "Osteichthyes." Bony fishes. Not monophyletic (sturgeons, paddlefish, herrings, carps, cods, trout, seahorses, tunas, lungfishes, and coelocanths).

TETRAPODS (SUBGROUP OF VERTEBRATES)

Class Amphibia. Amphibians. Require water to reproduce.
Order Caudata. Salamanders and newts.
Order Anura. Frogs, toads.
Order Apoda. Apodans (caecilians).

AMNIOTES (SUBGROUP OF TETRAPODS)

Class "Reptilia." Skin with scales, embryo protected and nutritionally supported by extraembryonic membranes.

Subclass Anapsida. Turtles, tortoises.
Subclass Lepidosaura. *Sphenodon*, lizards, snakes.
Subclass Archosaura. Crocodiles, alligators.

Class Aves. Birds. In some classifications birds are grouped in the archosaurs.

Order Struthioniformes. Ostriches.
Order Sphenisciformes. Penguins.
Order Procellariiformes. Albatrosses, petrels.
Order Ciconiiformes. Herons, bitterns, storks, flamingoes.
Order Anseriformes. Swans, geese, ducks.
Order Falconiformes. Eagles, hawks, vultures, falcons.
Order Galliformes. Ptarmigan, turkeys, domestic fowl.
Order Columbiformes. Pigeons, doves.
Order Strigiformes. Owls.
Order Apodiformes. Swifts, hummingbirds.
Order Passeriformes. Sparrows, jays, finches, crows, robins, starlings, wrens.
Order Piciformes. Woodpeckers, toucans.
Order Psittaciformes. Parrots, cockatoos, macaws.

Class Mammalia. Skin with hair; young nourished by milk-secreting mammary glands of adult.

Subclass Prototheria. Egg-laying mammals (monotremes; duckbilled platypus, spiny anteaters).

Subclass Metatheria. Pouched mammals or marsupials (opossums, kangaroos, wombats, Tasmanian devils).

Subclass Eutheria. Placental mammals.

Order Edentata. Anteaters, tree sloths, armadillos.
Order Insectivora. Tree shrews, moles, hedgehogs.
Order Chiroptera. Bats.
Order Scandentia. Insectivorous tree shrews.
Order Primates.

Suborder Strepsirhini (prosimians). Lemurs, lorises.
Suborder Haplorhini (tarsioids and anthropoids).

Infraorder Tarsiiformes. Tarsiers.
Infraorder Platyrrhini (New World monkeys).

Family Cebidae. Spider monkeys, howler monkeys, capuchin.

Infraorder Catarrhini (Old World monkeys and hominoids).

Superfamily Cercopithecoidea. Baboons, macaques, langurs.

Superfamily Hominoidea. Apes and humans.

Family Hylobatidae. Gibbon.

Family "Pongidae." Chimpanzees, gorillas, orangutans.

Family Hominidae. Existing and extinct human species (*Homo*) and humanlike species, including the australopiths.

Order Lagomorpha. Rabbits, hares, pikas.
Order Rodentia. Most gnawing animals (squirrels, rats, mice, guinea pigs, porcupines, beavers, etc.).
Order Carnivora. Carnivores (wolves, cats, bears, etc.).
Order Pinnipedia. Seals, walruses, sea lions.
Order Proboscidea. Elephants, mammoths (extinct).
Order Sirenia. Sea cows (manatees, dugongs).
Order Perissodactyla. Odd-toed ungulates (horses, tapirs, rhinos).
Order Tubulidentata. African aardvarks.
Order Artiodactyla. Even-toed ungulates (camels, deer, bison, sheep, goats, antelopes, giraffes, etc.).
Order Cetacea. Whales, porpoises.

Appendix II. Answers to Self-Quizzes

Italicized numbers refer to relevant section numbers

CHAPTER 17		**CHAPTER 18**		**CHAPTER 19**		**CHAPTER 20**	
1. d	*17.1*	1. populations	*18.1*	1. d	*19.1*	1. c	*20.1*
2. d	*17.4*	2. b	*Issues, Impacts*	2. d	*19.1*	2. c	*20.2*
3. a	*17.5*	3. a	*18.1*	3. a	*19.2*	3. c	*20.3*
4. Gondwana	*17.6*	4. c	*18.1, 18.3*	4. c	*19.4*	4. b	*20.4*
5. b	*17.7*	5. b	*18.4*	5. c	*19.4*	5. d	*20.4*
6. d	*17.7*	6. c	*18.5*	6. c	*19.5*	6. f	*20.1*
7. d	*17.8*	7. c	*18.6*	7. d	*19.5*	c	*20.2*
8. c	*17.8*	8. e	*18.6*	8. c	*19.6*	d	*20.3*
g	*17.4*	9. a	*18.7*	9. d	*19.5*	a	*20.3*
a	*17.4*	10. c	*18.8*	10. e	*19.5*	b	*20.5*
f	*17.8*	d	*18.1, 18.3*	d	*19.4*	e	*20.5*
e	*17.5*	a	*18.1*	a	*19.5*		
c	*17.7*	b	*18.7*	f	*19.5*		
b	*17.2*			b	*19.5, 19.6*		
d	*17.7*			c	*19.4*		

Appendix VI. Units of Measure

Metric-English Conversions

Length

English		Metric
inch	=	2.54 centimeters
foot	=	0.30 meter
yard	=	0.91 meter
mile (5,280 feet)	=	1.61 kilometer

To convert	multiply by	to obtain
inches	2.54	centimeters
feet	30.00	centimeters
centimeters	0.39	inches
millimeters	0.039	inches

Weight

English		Metric
grain	=	64.80 milligrams
ounce	=	28.35 grams
pound	=	453.60 grams
ton (short) (2,000 pounds)	=	0.91 metric ton

To convert	multiply by	to obtain
ounces	28.3	grams
pounds	453.6	grams
pounds	0.45	kilograms
grams	0.035	ounces
kilograms	2.2	pounds

Volume

English		Metric
cubic inch	=	16.39 cubic centimeters
cubic foot	=	0.03 cubic meter
cubic yard	=	0.765 cubic meters
ounce	=	0.03 liter
pint	=	0.47 liter
quart	=	0.95 liter
gallon	=	3.79 liters

To convert	multiply by	to obtain
fluid ounces	30.00	milliliters
quart	0.95	liters
milliliters	0.03	fluid ounces
liters	1.06	quarts

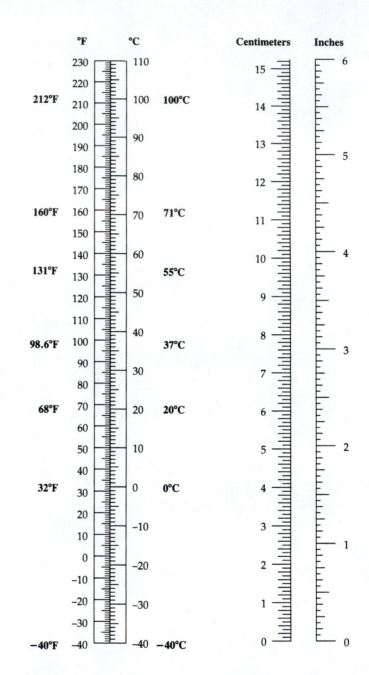

Glossary

adaptation, evolutionary [L. *adaptare*, to fit] Any long-term, heritable aspect of form, function, or behavior that improves an individual's chances of surviving and reproducing; outcome of natural selection and other microevolutionary processes.

adaptive radiation A macroevolutionary pattern. A burst of genetic divergences from a lineage that gives rise to many species, each able to use a novel resource or to move into a new, or newly vacated, habitat.

adaptive zone A set of different niches that become be filled by a group of species.

allele One of two or more molecular forms of a gene at a given locus; alleles arise by mutation and encode slightly different versions of the same trait.

allele frequency Abundance of one allele relative to others at a gene locus among individuals of a population.

allopatric speciation [Gk. *allos*, different, + L. *patria*, native land] Speciation model. A physical barrier arises and separates populations or subpopulations of a species, ends gene flow, and so favors divergences that result in new species.

anagenesis A major pattern of speciation. Directional changes in allele frequencies and morphology are confined within a single lineage, and in time a new type differs so much from the ancestral type that it is classified as a separate species.

analogous structures (an-AL-uh-gus) [Gk. *analogos*, similar to one another] Dissimilar body parts that have become similar in structure, function, or both in lineages that are not closely related but were subjected to similar pressures.

archipelago A chain or cluster of islands, often of volcanic origin in the open ocean.

balanced polymorphism An outcome of natural selection against homozygotes, so that two or more alleles for a trait are being maintained in the population.

big bang Model for the origin of universe, by a nearly instantaneous distribution of all matter and energy through all of space.

biogeography Scientific study of patterns in the geographic distribution of species and communities.

biological species concept Definition of a sexually reproducing species as one or more populations of individuals that interbreed under natural conditions, produce fertile offspring, and are reproductively isolated from other such populations.

bottleneck Severe reduction in the size of a population, brought about by intense selection pressure or a natural calamity.

catastrophism Idea that abrupt changes in the geologic and fossil records are evidence of divinely invoked catastrophes.

clade [Gk. *klados-*, branch] All species that share a unique trait, being descended from an ancestral species in which the trait first evolved.

cladogenesis One speciation pattern. A lineage branches when one or more of its populations or subpopulations become reproductively isolated, and then genetic divergences result in new species.

cladogram Evolutionary tree diagram that depicts relative relatedness among groups. Each branch is monopyletic; it includes only an ancestral species in which a unique trait first evolved and all of its descendants.

comparative morphology [Gk. *morph*, form] Scientific study of comparable external body parts of embryonic stages and adult forms of major lineages.

derived trait A novel feature shared only by descendants of an ancestral species in which it originated.

dimorphism Persistence of two forms of the same trait in a population.

directional selection Mode of natural selection by which forms at one end of a range of phenotypic variation are favored.

disruptive selection Mode of natural selection that favors different forms of a trait at both ends of a range of variation; intermediate forms are selected against.

endosymbiosis [*Endo–*, within + *symbiosis*, living together] An intimate, permanent ecological interaction in which one species lives and reproduces in the other's body to the benefit of one or both.

evolutionary tree A treelike diagram in which each branch point represents a divergence from a shared ancestor; each branch is a separate line of descent.

extinction Irrevocable loss of a species.

fitness The degree of adaptation to the environment, as measured by the relative genetic contribution to future generations.

fixation Of a population, the loss of all alleles but one at a gene locus; all individuals have become homozygous for the allele.

fossil Recognizable, physical evidence of an organism that lived in the distant past.

fossilization How fossils form over time. An organism or evidence of it gets buried in sediments or volcanic ash; water slowly infiltrates the remains, and metal ions and

other inorganic compounds dissolved in it replace the minerals in bones and other hardened tissues.

founder effect A form of bottlenecking. By chance, a few individuals that establish a new population differ in allele frequencies relative to the original population.

gene flow Microevolutionary process; alleles enter and leave a population by immigration and emigration. Counters mutation, natural selection, and genetic drift, hence reproductive isolation.

gene pool All genotypes in a population; a pool of genetic resources.

genetic divergence An accumulation of differences in the gene pools of two or more populations or subpopulations of a species after gene flow stops entirely; mutation, natural selection, and genetic drift operate independently in each one.

genetic drift Change in allele frequencies over generations due to chance alone. Most pronounced effects in small populations.

genetic equilibrium In theory, a state in which a population is not evolving with respect to a specified gene locus. *Compare* Hardy–Weinberg rule.

geologic time scale Time scale for Earth's history; major subdivisions correspond to mass extinctions. Dates are now absolute as a result of radiometrically dating.

Gondwana Paleozoic supercontinent that later became part of Pangea.

gradual model, speciation Addresses the rate of speciation and cites fossil evidence that morphological changes accumulate slowly over great time spans.

half-life The unvarying time it takes for half of a quantity of any radioisotope to decay into a more stable form.

homologous structures Of separate lineages, comparable body parts that show underlying similarity even when they may differ in size, shape, or function; outcome of morphological divergence from a shared ancestor.

inbreeding Nonrandom mating among very close relatives that share many identical alleles; may fix harmful alleles.

key innovation A chance modification in some body structure or function that gives a species the opportunity to exploit the environment more efficiently or in a novel way; e.g., modifications of the forelimbs of amniotes into diverse legs and wings during radiations into adaptive zones.

lethal mutation Mutation having drastic effects on phenotype; usually causes death.

macroevolution Large-scale patterns, rates of change, and trends among lineages.

mass extinction Catastrophic event or phase in geologic time when families or other major groups are lost.

microevolution Of a population, a small-scale change in allele frequencies resulting from mutation, genetic drift, gene flow, natural selection, or a combination of them.

molecular clock Model used to calculate the time of origin of one lineage relative to others; assumes that a group of genes accumulates mutations at a constant rate, measurable as a series of predictable ticks back through time. The last tick stops close to the time the lineage originated.

monophyletic group A set of species that share a derived trait, a novel feature that evolved in one species and is present only in its descendants; all of the evolutionary branchings from a single stem.

morphological convergence A pattern of macroevolution. In response to similar environmental pressures, body parts of evolutionarily distant lineages slowly evolve in similar ways and end up being alike in function, appearance, or both.

morphological divergence Pattern of macroevolution. One or more body parts of genetically diverging lineages undergo structural and functional changes from the parts in the common ancestor.

natural selection Microevolutionary process; the outcome of differences in survival and reproduction among individuals of a population that differ in the details of their heritable traits.

neutral mutation A mutation with no effect on phenotype; natural selection thus cannot change its frequency in a population.

nucleic acid hybridization Any base-pairing between DNA or RNA strands from different sources.

Pangea Paleozoic supercontinent; the first land plants and animals evolved on it.

parapatric speciation A speciation model. Populations in contact along a common border evolve into new species; hybrids that form in the contact zone are less fit than individuals on either side of it and thereby act as a reproductive isolating mechanism.

plate tectonics Theory that great slabs or plates of Earth's outer layer float on a hot, semi-molten mantle. All plates are moving slowly and have rafted continents to new positions over time.

polymorphism (poly-MORE-fizz-um) [Gk. *polus*, many, + *morphe*, form] Persistence of two or more qualitatively different forms of a trait, or morphs, in a population.

polyploidy A case of somatic cells having three or more of each type of chromosome characteristic of the species.

population All individuals of the same species living in a specified area.

proto-cell Presumed stage of chemical evolution that preceded living cells.

punctuation model, speciation Addresses the rate of speciation; cites fossil evidence that morphological changes required for reproductive isolation evolve in a relatively brief time span, within the tens to hundreds of thousands of years when two or more populations are diverging from each other.

radiometric dating Method of measuring proportions of a radioisotope in a mineral trapped long ago in newly formed rock and a daughter isotope that formed from it by radioactive decay in the same rock. Used to assign absolute dates to fossil-containing rocks and to the geologic time scale.

reproductive isolating mechanism Any heritable feature of body form, function, or behavior that prevents interbreeding between two or more populations; sets the stage for genetic divergences.

RNA world Model for a time prior to the evolution of DNA; a self-replicating system chemically evolved in which RNA strands were templates for protein synthesis.

sexual dimorphism A notable difference between female and male phenotypes of a population.

sexual selection A category of natural selection; an outcome of differences in success at attracting mates and reproducing among individuals of a population.

six-kingdom classification system The grouping of all organisms into kingdoms Bacteria, Archaea, Protista, Fungi, Plantae, and Animalia.

speciation (spee-see-AY-shun) One of the macroevolutionary processes; formation of daughter species from a population or subpopulation of a parent species; the routes vary in their details and duration.

species (SPEE-sheez) [L. *species*, a kind] Of sexually reproducing species, one or more natural populations of individuals that successfully interbreed and are isolated reproductively from other such groups. By a cladistic definition, one or more natural populations of individuals with at least one unique trait derived a common ancestor that occurs in no other groups.

stabilizing selection Mode of natural selection; intermediate phenotypes are favored over extremes at both ends of the range of variation.

stratification Stacks of sedimentary rock layers, built up by deposition of silt and other materials over time.

stromatolite Fossilized remains of dome-shaped mats of shallow-water communities, cyanobacterial species especially, that were infiltrated with dissolved minerals and fine sediments. Some are 3 billion years old.

sympatric speciation [Gk. *sym*, together, + *patria*, native land] A speciation model. Occurs inside the home range of a species in the absence of a physical barrier; e.g., by way of polyploidy in flowering plants.

taxon, plural **taxon** A set of organisms of a given type.

three-domain system A classification system that groups all organisms into domains Bacteria, Archaea, and Eukarya.

uniformity theory Theory that Earth's surface has changed in slow, uniformly repetitive ways except for expected annual catastrophes, such as big floods. Changed Darwin's view of evolution; has since been discredited by plate tectonics theory.

Art Credits and Acknowledgments

This page constitutes an extension of the book copyright page. We have made every effort to trace the ownership of all copyrighted material and secure permission from copyright holders. In the event of any question arising as to the use of any material, we will be pleased to make the necessary corrections in future printings. Thanks are due to the following authors, publishers, and agents for permission to use the material indicated.

Page iii, v Photographer Russ Lowgren.

TABLE OF CONTENTS **Page xi** Christopher Ralling. **Page viii** From left, © Francois Gohier/Photo Researchers, Inc.; © David Parker/SPL/Photo Researchers, Inc.; © Jack Jeffrey Photography.

INTRODUCTION NASA Space Flight Center

Page 259 UNIT III © Wolfgang Kaehler/Corbis.

CHAPTER 17 **17.1** Left, NASA Galileo Imaging Team; Right © David A. Kring, NASA/Univ. Arizona Space Imagery Center. **17.2** Art by Don Davis. **17.3** (a, c) © Wolfgang Kaehler/Corbis; (b) © Earl & Nazima Kowall/Corbis; (d, e) Edward S. Ross. **17.4** right, © Bruce J. Mohn; (inset) Phillip Gingerich, Director, University of Michigan. Museum of Paleontology. **17.5** © Jonathan Blair/Corbis. **17.6** (a) Courtesy George P. Darwin, Darwin Museum, Down House; (b) Christopher Ralling; (e) Dieter & Mary Plage/Survival Anglia. **Page 265** Heather Angel. **17.7** (a) © Joe McDonald/Corbis; (b) © Karen Carr Studio/www.karencarr.com. **17.8** (a) © Gerra and Sommazzi/www.justbirds.org; (b) © Kevin Schafer/Corbis; (c) Alan Root/Bruce Coleman Ltd. **17.9** Down House and The Royal College of Surgeons of England. **17.10** Left, H. P. Banks; Right, Jonathan Blair. **17.11** © Jonathan Blair/Corbis. **17.12** (b) © 2001 Photodisc, Inc. **17.14** © Corbis. **17.15** (a) NASA/GSFC. **17.16** (a–d) After A.M. Ziegler, C.R. Scotese, and S.F. Barrett, "Mesozoic and Cenozoic Paleogeographic Maps," and J. Krohn and J. Sundermann (Eds.), Tidal Frictions and the Earth's Rotation II, Springer-Verlag, 1983; (e) © Martin Land/Photo Researchers, Inc.; (f) © John Sibbick. **17.18** (a) J. Scott Altenbach, University of New Mexico, computer enhanced by Lisa Starr; (b) Frans Lanting/Minden Pictures; computer enhanced by Lisa Starr; (c) above, © Stephen Dalton/Photo Researchers, Inc.); (c bottom) Natural History Collection, Royal BC Museum. **17.19** (a, b) Courtesy of Professor Richard Amasino,

University of Wisconsin-Madison; (c) Juergen Berger, Max Planck Institute for Developmental Biology—Tuebingen, Germany; (d) Courtesy of Professor Martin F. Yanofsky, UCSD. **17.21** (a) above, Tait/Sunnucks Peripatus Research; (a) below, © Jennifer Grenier, Grace Boekhoff-Falk and Sean Carroll, HMI, University of Wisconsin-Madison; (b) above, Herve Chaumeton/Agence Nature; (b) below, © Jennifer Grenier, Grace Boekhoff-Falk and Sean Carroll, HMI, University of Wisconsin-Madison; (c) above, © Peter Skinner/Photo Researchers, Inc.; (c) below, Courtesy of Dr. Giovanni Levi; (d) Dr. Chip Clark. **Page 278** © TEK IMAGE/Photo Researchers, Inc. **17.23** Left, Kjell B. Sandved/Visuals Unlimited; Center, Jeffrey Sylvester/TPG/Getty Images; Right, Thomas D. Mangelsen/Images of Nature. **17.24** John Klausmeyer, University of Michigan Exhibit of Natural History.

CHAPER 18 **18.1** Above, © Bettmann/Corbis; Below, © Reuters NewMedia, Inc./Corbis. **Page 283** © St. Bartholomew's Hospital/Science Photo Library/Photo Researchers, Inc. **18.2** Clockwise from left, © Peter Bowater/Photo Researchers, Inc.; © Owen Franken/Corbis; © Sam Kleinman/Corbis; Alan Solem; © Christopher Briscoe/Photo Researchers, Inc.; © Jim Cornfield/Corbis. **18.3** © 2002/Photodisc/Getty Images. **18.6** J. A. Bishop, L. M. Cook. **18.7** Courtesy of Hopi Hoekstra, University of California, San Diego. **18.10** © Peter Chadwick/Science Photo Library/Photo Researchers, Inc.. **18.11** Thomas Bates Smith. **18.12** Bruce Beehler. **18.13** (a–b) After Ayala and others; (c) © Michael Freeman/Corbis. **18.14** Adapted from S. S. Rich, A. E. Bell, and S. P. Wilson, "Genetic drift in small populations of Tribolium," *Evolution* 33:579-584, Fig. 1, p. 580, 1979. Used by permission of the publisher. **Page 294** © Steve Bronstein/The Image Bank/Getty Images. **18.15** Frans Lanting/Minden Pictures (computer-modified by Lisa Starr). **18.16** Left, David Neal Parks; right W. Carter Johnson. **18.17** (b) John W. Merck, Jr., University of Maryland. **18.18** Left, © Thomas Mangelsen; right © Theo Allofs/Corbis. **18.19** Left, © Francois Gohier/Photo Researchers, Inc.; right © David Parker/SPL/Photo Researchers, Inc.. **18.20** Elliot Erwitt/Magnum Photos, Inc..

CHAPTER 19 **19.1** Left, © Jack Jeffrey Photography; Right, Image courtesy of the Image Analysis Laboratory, NASA/Johnson Space Center. **19.2** Bill Sparklin/Ashley Dayer. **19.4** (a) John Alcock, Arizona State University; (b) © Alvin E.

Staffan/Photo Researchers, Inc.; (c) G. Ziesler/ZEFA. **19.5** Left © Digital Vision/PictureQuest; (a) © Joe McDonald/Corbis. **19.6** (a) © Graham Neden/Corbis; (b) © Kevin Schafer/ Corbis; Center, © Ron Blakey, Northern Arizona University; (c) © Rick Rosen/Corbis SABA. **19.7** Po' ouli, Bill Sparklin/Ashley Dayer; All others, © Jack Jeffrey Photography. **19.8** Above, Steve Gartlan; Below, © Below Water Photography/www.belowwater.com. **19.9** Jean-Claude Carton/Bruce Coleman, Inc. **19.10** After W. Jensen and F. B. Salisbury, Botany: An Ecological Approach, Wadsworth, 1972. **19.11** Courtesy of Dr. Robert Mesibov. **19.13** Courtesy of Daniel C. Kelley, Anthony J. Arnold, and William C. Parker, Florida State University Department of Geological Science. **19.14** © Carnegie Museum of Natural History. **19.15** From left, © Science Photo Library/Photo Researchers, Inc.; © Galen Rowell/Corbis; © Kevin Schafer/Corbis; Courtesy of Department of Entomology, University of Nebraska-Lincoln; Bruce Coleman, Ltd. **19.18** From left, © Hans Reinhard/Bruce Coleman, Inc; © Phillip Colla Photography; © Randy Wells/Corbis; © Cousteau Society/The Image Bank/Getty Images; © Robert Dowling/Corbis. **19.20** Left, Courtesy of Department of Library Services, American Museum of Natural History (Neg. #K10273); Right, Photo by Lisa Starr. **19.21** © Gulf News, Dubai, UAE.

CHAPTER 20 **20.1** Left, Courtesy of Agriculture Canada; Right, © Raymond Gehman/Corbis. **20.2** © Philippa Uwins/The University of Queensland. **20.3** Jeff Hester and Paul Scowen, Arizona State University, and NASA. **20.4** Left, Painting by William K. Hartmann; Right, Painting by Chesley Bonestell. **20.6** (a) Eiichi Kurasawa/Photo Researchers, Inc.; (b) © Dr. Ken MacDonald/SPL/Photo Researchers, Inc.; (c) © Micheal J. Russell, Scottish Universities Environmental Research Centre. **20.7** (a) Sidney W. Fox; (b) From Hanczyc, Fujikawa, and Szostak, Experimental Models of Primitive Cellular Compartments: Encapsulation, Growth, and Division; www.sciencemag.org, *Science* 24 October 2003; 302;529, Figure 2, page 619. Reprinted with Permission of the authors and AAAS. **20.8** (a) Stanley M. Awramik; (b-c) © Bruce Runnegar, NASA Astrobiology Institute; (d) © N. J. Butterfield, University of Cambridge. **20.9** (a) © Chase Studios/Photo Researchers, Inc.; (b) © John Reader/SPL/Photo Researchers, Inc.; (c) © Sinclair Stammers/SPL/Photo Researchers, Inc. **20.10** (a) © CNRI/Photo Researchers, Inc.; (b) © Robert Trench, Professor Emeritus, University of British Columbia.

Index

The letter i designates illustration; t designates table; **bold** designates defined term;
■ highlights the location of applications contained in text.